ALSO BY JOHN BROCKMAN

AS AUTHOR
By the Late John Brockman
37
Afterwords
The Third Culture: Beyond the Scientific Revolution
Digerati

AS EDITOR
About Bateson
Speculations
Doing Science
Ways of Knowing
Creativity
The Greatest Inventions of the Past 2,000 Years
The Next Fifty Years
The New Humanists
Curious Minds
What We Believe but Cannot Prove
My Einstein
Intelligent Thought
What Is Your Dangerous Idea?
What Are You Optimistic About?
What Have You Changed Your Mind About?
This Will Change Everything
Is the Internet Changing the Way You Think?
Culture
The Mind
This Will Make You Smarter

AS COEDITOR
How Things Are (with Katinka Matson)

HARPER ●❍ PERENNIAL

NEW YORK ● LONDON ● TORONTO ● SYDNEY ● NEW DELHI ● AUCKLAND

This

Deep, Beautiful, and Elegant Theories of How the World Works

Explains

Edited by
JOHN BROCKMAN

Everything

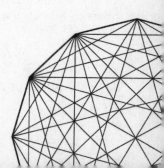

FIRST EDITION

Library of Congress Cataloging-in-Publication Data
This explains everything : deep, beautiful, and elegant theories of how the world works / edited by John Brockman. — 1st ed.
    p. cm.
 Summary: "Drawn from the cutting-edge frontiers of science, *This Explains Everything* presents 150 of the most deep, surprising, and brilliant explanations of how the world works, with contributions by Jared Diamond, Richard Dawkins, Nassim Taleb, Brian Eno, Steven Pinker, and more"— Provided by publisher.
  ISBN 978-0-06-223017-1 (pbk.) — ISBN 0-06-223017-4 (paperback) 1. Science—Miscellanea. 2. Explanation. I. Brockman, John, 1941–
 Q173.T53 2013
 500—dc23
                              2012032107

13 14 15 16 17   OV/RRD   10 9 8 7 6 5 4

CONTENTS

ACKNOWLEDGMENTS

I wish to thank Peter Hubbard of HarperCollins for his encouragement. I am also indebted to my agent, Max Brockman, who saw the potential for this book, and to Sara Lippincott for her thoughtful and meticulous editing.

PREFACE

THE *EDGE* QUESTION

In 1981, I founded the Reality Club. From its founding through 1996, the club held its meetings in Chinese restaurants, artists' lofts, the boardrooms of investment-banking firms, ballrooms, museums, and living rooms, among other venues. The Reality Club differed from the Algonquin Round Table, the Apostles, and the Bloomsbury Group, but it offered the same quality of intellectual adventure. Perhaps the closest resemblance was to the late 18th- and early 19th-century Lunar Society of Birmingham, an informal gathering of the leading cultural figures of the new industrial age—James Watt, Erasmus Darwin, Josiah Wedgwood, Joseph Priestley, Benjamin Franklin. In a similar fashion, the Reality Club was an attempt to gather together those people exploring the themes of the post–Industrial Age.

In 1997, the Reality Club went online, rebranded as *Edge*. The ideas presented on *Edge* are speculative; they represent the frontiers in such areas as evolutionary biology, genetics, computer science, neurophysiology, psychology, cosmology, and physics. Emerging out of these contributions is a new natural philosophy, new ways of understanding physical systems, new ways of thinking that call into question many of our basic assumptions.

For each of the anniversary editions of *Edge*, I and a number of *Edge* stalwarts, including Stewart Brand, Kevin Kelly, and George Dyson, get together to plan the annual *Edge* Question—usually one that comes to one or another of us or our correspondents in the middle of the night. It's not easy coming up with a question. (As the late James Lee Byars, my friend and sometime collaborator, used to say: "I can answer the question, but am I bright enough to ask it?") We look for questions that inspire unpredictable answers—that provoke people into thinking thoughts they normally might not have. For this year's question, our thanks go, once again, to Steven Pinker.

Perhaps the greatest pleasure in science comes from theories that derive the solution to some deep puzzle from a small set of simple principles in a surprising way. These explanations are called "beautiful" or "elegant." Historical examples are Kepler's explanation of complex planetary motions as simple ellipses, Niels Bohr's explanation of the periodic table of the elements in terms of electron shells, and James Watson and Francis Crick's explanation of genetic replication via the double helix. The great theoretical physicist P. A. M. Dirac famously said that "it is more important to have beauty in one's equations than to have them fit experiment."

The *Edge* Question 2012
WHAT IS YOUR FAVORITE DEEP, ELEGANT, OR BEAUTIFUL EXPLANATION?

The online response to the *Edge* website this year (http://edge.org/annual-question/) was enormous—some 200 provocative (and often lengthy) discussions. What follows is necessarily an edited selection. In the spirit of *Edge*, the contributions presented here embrace scientific thinking in the broadest sense: as the most reliable way of gaining knowledge about anything—including such fields of inquiry as philosophy, mathematics, economics, history, language, and human behavior. The common thread is that a simple and nonobvious idea is proposed as the explanation for a diverse and complicated set of phenomena.

JOHN BROCKMAN
Publisher & Editor, Edge

EVOLUTION BY MEANS OF NATURAL SELECTION

SUSAN BLACKMORE

Psychologist; author, Consciousness: An Introduction

Of course it has to be Darwin. Nothing else comes close. Evolution by means of natural selection (or indeed any kind of selection—natural or unnatural) provides the most beautiful, elegant explanation in all of science. This simple three-step algorithm explains, with one simple idea, why we live in a universe full of design. It explains not only why we are here but why trees, kittens, Urdu, the Bank of England, Chelsea football team, and the iPhone are here.

You might wonder why, if this explanation is so simple and powerful, no one thought of it before Darwin and Alfred Russel Wallace did, and why even today so many people fail to grasp it. The reason, I think, is that at its heart there seems to be a tautology. It seems as though you are saying nothing when you say that "Things that survive survive" or "Successful ideas are successful." To turn these tautologies into power, you need to add the context of a limited world in which not everything survives and competition is rife, and also realize that this is an ever-changing world in which the rules of the competition keep shifting.

In that context, being successful is fleeting, and now the three-step algorithm can turn tautology into deep and elegant explanation. Copy the survivors many times with slight variations and let them loose in this ever-shifting world, and only those suited to the new conditions will carry on. The world fills with creatures, ideas, institutions, languages, stories, software, and machines that have all been designed by the stress of this competition.

This beautiful idea *is* hard to grasp, and I have known many university students who have been taught evolution at school and thought they understood it, but have never really done so. One of the joys of teaching, for me, was to see that astonished look on students' faces when they suddenly got it. That was heartwarming indeed. But I also call it heartwarming because, unlike some religious folk, when I look out of my window past my computer to the bridge over the river and the trees and cows in the distance, I delight in the simple and elegant competitive process that brought them all into being, and at my own tiny place within it all.

LIFE IS A DIGITAL CODE

MATT RIDLEY

Science writer; founding chairman, International Centre for Life; author, The Rational Optimist

It's hard now to recall just how mysterious life was on the morning of February 28 and just how much that had changed by lunchtime. Look back at all the answers to the question "What is life?" from before that, and you get a taste of just how we, as a species, floundered. Life consisted of three-dimensional objects of specificity and complexity (mainly proteins). And it copied itself with accuracy. How? How do you set about making a copy of a three-dimensional object? How do you grow it and develop it in a predictable way? This is the one scientific question whose answer absolutely nobody came close to guessing. Erwin Schrödinger had a stab but fell back on quantum mechanics, which was irrelevant. True, he used the phrase "aperiodic crystal," and if you are generous you can see that as a prediction of a linear code, but I think that's stretching generosity.

Indeed, the problem had just got even more baffling, thanks to the realization that DNA played a crucial role—because DNA was monotonously simple. All the explanations of life before February 28, 1953, are handwaving waffle and might as well have spoken of protoplasm and vital sparks for all the insight they gave.

Then came the double helix, and the immediate understanding that, as Francis Crick wrote to his son a few weeks later, "some sort of code"—digital, linear, two-dimensional, combinatorially infinite, and instantly self-replicating—was all the explanation you needed. Here's part of Crick's letter, March 17, 1953:

My Dear Michael,

Jim Watson and I have probably made a most important discovery. . . . Now we believe that the DNA is a code. That is, the order of the bases (the letters) makes one gene different from another gene (just as one page of print is different from another). You can see how Nature makes copies of the genes. Because if the two chains unwind into two separate chains, and if each chain makes another chain come together on it, then because A always goes with T, and G with C, we shall get two copies where we had one before. In other words, we think we have found the basic copying mechanism by which life comes from life. . . . You can understand we are excited.

Never has a mystery seemed more baffling in the morning and an explanation more obvious in the afternoon.

REDUNDANCY REDUCTION
PATTERN RECOGNITION

RICHARD DAWKINS
Evolutionary biologist; Emeritus Professor of the Public
Understanding of Science, Oxford; author, The Magic of Reality

Deep, elegant, beautiful? Part of what makes a theory elegant is its power to explain much while assuming little. Here, Darwin's natural selection wins hands down. The ratio of the huge amount that it explains (everything about life: its complexity, diversity, and illusion of crafted design) divided by the little that it needs to postulate (nonrandom survival of randomly varying genes through geological time) is gigantic. Never in the field of human comprehension were so many facts explained by assuming so few. Elegant then, and deep—its depths hidden from everybody until as late as the 19th century. On the other hand, for some tastes, natural selection is too destructive, too wasteful, too cruel to count as beautiful. In any case, I can count on somebody else choosing Darwin. I'll take his great-grandson instead, and come back to Darwin at the end.

Horace Barlow, FRS, is the youngest grandchild of Sir Horace Darwin, Charles Darwin's youngest child. Now a very active ninety, Barlow is a member of a distinguished lineage of Cambridge neurobiologists. I want to talk about an idea he published in two papers in 1961, on redundancy reduction and pattern recognition. It's an idea whose ramifications and significance have inspired me throughout my career.

The folklore of neurobiology includes a mythical "grandmother neuron," which fires only when a very particular image, the face

Lettvin's grandmother, falls on the retina (Lettvin was a
distinguished American neurobiologist who, like Barlow, worked
on the frog retina). The point is that Lettvin's grandmother is
only one of countless images that a brain is capable of recognizing.
If there were a specific neuron for everything we can recognize—
not just Lettvin's grandmother but lots of other faces, objects, letters of the alphabet, flowers, each one seen from many angles and
distances—we would have a combinatorial explosion. If sensory
recognition worked on the grandmother principle, the number
of specific-recognition neurons for all possible combinations of
nerve impulses would exceed the number of atoms in the universe.
Independently, the American psychologist Fred Attneave had cal-
culated that the volume of the brain would have to be measured
in cubic light-years. Barlow and Attneave independently proposed
redundancy reduction as the answer.

Claude Shannon, inventor of information theory, coined
"redundancy" as a kind of inverse of information. In English, "q"
is always followed by "u," so the "u" can be omitted without loss
of information. It is redundant. Wherever redundancy occurs in a
message (which is wherever there is nonrandomness), the message
can be more economically recoded without loss of information—
although with some loss in capacity to correct errors. Barlow
suggested that at every stage in sensory pathways there are mechanisms tuned to eliminate massive redundancy.

The world at time *t* is not greatly different from the world at
time *t*-1. Therefore it is not necessary for sensory systems continuously to report the state of the world. They need only signal
changes, leaving the brain to assume that everything not reported
remains the same. Sensory adaptation is a well-known feature of
sensory systems, which does precisely as Barlow prescribed. If a
neuron is signaling temperature, for example, the rate of firing is
not, as one might naively suppose, proportional to the tempera-

ture. Instead, firing rate increases only when there is a *change* in temperature. It then dies away to a low, resting frequency. The same is true of neurons signaling brightness, loudness, pressure, and so on. Sensory adaptation achieves huge economies by exploiting the nonrandomness in temporal sequence of states of the world.

What sensory adaptation achieves in the temporal domain, the well-established phenomenon of lateral inhibition does in the spatial domain. If a scene in the world falls on a pixelated screen, such as the back of a digital camera or the retina of an eye, most pixels seem the same as their immediate neighbors. The exceptions are those pixels which lie on edges, boundaries. If every retinal cell faithfully reported its light value to the brain, the brain would be bombarded with a hugely redundant message. Great economies can be achieved if most of the impulses reaching the brain come from pixel cells lying along edges in the scene. The brain then assumes uniformity in the spaces between edges.

As Barlow pointed out, this is exactly what lateral inhibition achieves. In the frog retina, for example, every ganglion cell sends signals to the brain, reporting on the light intensity in its particular location on the surface of the retina. But it simultaneously sends inhibitory signals to its immediate neighbors. This means that the only ganglion cells to send strong signals to the brain are those that lie on an edge. Ganglion cells lying in uniform fields of color (the majority) send few if any impulses to the brain, because they, unlike cells on edges, are inhibited by all their neighbors. The spatial redundancy in the signal is eliminated.

The Barlow analysis can be extended to most of what is now known about sensory neurobiology, including Hubel and Wiesel's famous horizontal- and vertical-line detector neurons in cats (straight lines are redundant, reconstructable from their ends), and in the movement ("bug") detectors in the frog retina, discovered by the same Jerry Lettvin and his colleagues. Movement represents a nonredundant change in the frog's world. But even movement is redundant if it persists in the same direction at the same speed. Sure enough, Lettvin and colleagues discovered a "strangeness" neuron in their frogs, which fires only when a moving object does something unexpected, such as speeding up, slowing down, or changing direction. The strangeness neuron is tuned to filter out redundancy of a very high order.

Barlow pointed out that a survey of the sensory filters of a given animal could, in theory, give us a readout of the redundancies present in the animal's world. They would constitute a kind of description of the statistical properties of that world. Which reminds me, I said I'd return to Darwin. In *Unweaving the Rainbow*, I suggested that the gene pool of a species is a "Genetic Book of the Dead," a coded description of the ancestral worlds in which the genes of the species have survived through geological time. Natural selection is an averaging computer, detecting redundancies—repeat patterns—in successive worlds (successive through millions of generations) in which the species has survived (averaged over all members of the sexually reproducing species). Could we take what Barlow did for neurons in sensory systems and do a parallel analysis for genes in naturally selected gene pools? Now, that would be deep, elegant, *and* beautiful.

THE POWER OF ABSURDITY

SCOTT ATRAN

Anthropologist, Centre National de la Recherche Scientifique, Paris; author, Talking to the Enemy: Faith, Brotherhood, and the (Un)Making of Terrorists

The notion of a transcendent force that moves the universe or history or determines what is right and good—and whose existence is fundamentally beyond reason and immune to logical or empirical disproof—is the simplest, most elegant, and most scientifically baffling phenomenon I know of. Its power and absurdity perturbs mightily and merits careful scientific scrutiny. In an age in which many of the most volatile and seemingly intractable conflicts stem from sacred causes, scientific understanding of how best to deal with the subject has also never been more crucial.

Call it love of Group or God, or devotion to an Idea or Cause, it matters little in the end. It is the "the privilege of absurdity; to which no living creature is subject, but man only," of which Hobbes wrote in *Leviathan*. In *The Descent of Man*, Darwin cast it as the virtue of "morality," with which winning tribes are better endowed in history's spiraling competition for survival and dominance. Unlike other creatures, humans define the groups they belong to in abstract terms. Often they strive to achieve a lasting intellectual and emotional bond with anonymous others and seek to heroically kill and die not in order to preserve their own lives or those of people they know but for the sake of an idea—the conception they have formed of themselves, of "who we are."

Sacred, or transcendental, values and religious ideas are culturally universal, yet content varies markedly across cultures. Sacred values mark the moral boundaries of societies and determine which material transactions are permissible. Material transgressions of the sacred are taboo: We consider people who sell their children or sell out their country to be sociopaths; other societies consider adultery or disregard of the poor immoral, but not necessarily selling children or women or denying freedom of expression.

Sacred values usually become strongly relevant only when challenged, much as food takes on overwhelming value in people's lives only when denied. People in one cultural milieu are often unaware of what is sacred for another—or, in becoming aware through conflict, find the other side's values (pro-life vs. pro-choice, say) immoral and absurd. Such conflicts cannot be wholly reduced to secular calculations of interest but must be dealt with on their own terms, a logic different from the marketplace or *realpolitik*. For example, cross-cultural evidence indicates that the prospect of crippling economic burdens and huge numbers of deaths doesn't necessarily sway people from choosing to go to war, or to opt for revolution or resistance. As Darwin noted, the virtuous and brave do what is right, regardless of consequences, as a moral imperative. (Indeed, we have suggestive neuroimaging evidence that people process sacred values in parts of the brain devoted to rule-bound behavior rather than utilitarian calculations—think Ten Commandments or Bill of Rights.)

There is an apparent paradox underlying the formation of large-scale human societies. The religious and ideological rise of civilizations—of larger and larger agglomerations of genetic strangers, including today's nations, transnational movements, and other "imagined communities" of fictive kin—seem to depend upon what Kierkegaard deemed this "power of the preposterous" (as in Abraham's willingness to slit the throat of his most beloved son to show commitment to an invisible, no-name deity, thus making him the world's greatest culture hero rather than a child abuser, would-be murderer, or psychotic). Humankind's strongest social bonds and actions, including the capacities for cooperation and forgiveness, and for killing and allowing oneself to be killed, are born of commitment to causes and courses of action that are "ineffable"—that is, fundamentally immune to logical assessment for consistency and to empirical evaluation for costs and consequences. The more materially inexplicable one's devotion and commitment to a sacred cause—that is, the more absurd—the greater the trust others place in it and the more that trust generates commitment on their part.

To be sure, thinkers of all persuasions have tried to explain the paradox (most being ideologically motivated and simpleminded), often to show that religion is good, or more usually that religion is unreasonably bad. If anything, evolution teaches that humans are creatures of passion and that reason itself is primarily aimed at social victory and political persuasion rather than philosophical or scientific truth. To insist that persistent rationality is the best means and hope for victory over enduring irrationality—that logical harnessing of facts could someday do away with the sacred and so end conflict—defies all that science teaches about our passion-driven nature. Throughout the history of our species, as for the most intractable conflicts and greatest collective expressions of joy today, utilitarian logic is a pale prospect to replace the sacred.

For Alfred Russel Wallace, moral behavior (along with mathematics, music, and art) was evidence that humans had not evolved through natural selection alone: "The special faculties we have been discussing clearly point to the existence in man of something which he has not derived from his animal progenitors—something which we may best refer to as being of a spiritual essence . . . beyond all explanation by matter, its laws and forces."* His disagreement with Darwin on this subject was longstand-

---

*   Alfred Russel Wallace, *Darwinism* (New York: Macmillan, 1889), 474–475.

ing, at one point prompting the latter to protest, "I hope you have not murdered too completely your own and my child."* But Darwin himself produced no causal account of how humans became moral animals, other than to say that because our ancestors were so physically weak, only group strength could get them through. Religion and the sacred, banned so long from reasoned inquiry by the ideological bias of all persuasions—perhaps because the subject is so close to who we want or don't want to be—is still a vast, tangled, and largely unexplored domain for science, however simple and elegant for most people everywhere in everyday life.

---

* Charles Darwin to A. R. Wallace, March 27, 1869. *Alfred Russel Wallace: Letters & Reminiscences*, ed. J. Marchant (New York: Harper, 1916), 197.

HOW APPARENT FINALITY CAN EMERGE

CARLO ROVELLI
*Theoretical physicist, Centre de Physique Théorique,
University of Marseille; author,* Quantum Gravity

Darwin, no doubt. The beauty and the simplicity of his explanation is astonishing. I am sure that others have pointed out Darwinian natural selection as their favorite deep, elegant, beautiful explanation, but I still want to emphasize the general reach of Darwin's central intuition, which goes well beyond the monumental result of having clarified that we share the same ancestors with all living beings on Earth and is directly relevant to the core of the entire scientific enterprise.

Shortly after the ancient Greek physicists started developing naturalistic explanations of nature, a general objection arose. The objection is well articulated in Plato—for instance, in the *Phaedo*—and especially in Aristotle's discussion of the theory of the "causes." Naturalistic explanations rely on what Aristotle called "the efficient cause"—namely, past phenomena producing effects. But the world appears to be dominated by phenomena that can be understood in terms of "final causes"—that is, an "aim" or a "purpose." These are evident in the kingdom of life. We have mouths "so" we can eat. The importance of this objection cannot be underestimated. It brought down ancient naturalism, and in the minds of many it is still the principal source of psychological resistance to a naturalistic understanding of the world.

Darwin discovered the spectacularly simple mechanism by which efficient causes produce phenomena that appear to be governed by final causes. Anytime we have phenomena that can reproduce, the actual phenomena we observe are those that keep reproducing and therefore are necessarily better at reproducing, and we can thus read them in terms of final causes. In other words, a final cause can be effective for understanding the world because it's a shortcut in accounting for the past history of a continuing phenomenon.

To be sure, this idea has appeared before. Empedocles speculated that the apparent finality in the living kingdom could be the result of selected randomness, and Aristotle himself, in his *Physics*, mentions a version of this idea for species ("seeds"). But the times were not yet ripe and the suggestion was lost in the following religious ages. I think the resistance to Darwin is not just difficulty in seeing the power of a spectacularly beautiful explanation but fear of realizing the extraordinary power such an explanation has in shattering old worldviews.

THE OVERDUE DEMISE OF MONOGAMY

AUBREY DE GREY
Gerontologist; chief science officer, SENS Foundation; author, Ending Aging

There are many persuasive arguments from evolutionary biology explaining why various species, notably *Homo sapiens*, have adopted a lifestyle in which males and females pair up long-term. But my topic here is not one of those explanations. Instead, it is the explanation for why we are close—far closer than most people, even most readers of *Edge*, yet appreciate—to the greatest societal, as opposed to technological, advance in the history of civilization.

In 1971, the American philosopher John Rawls coined the term "reflective equilibrium" to denote "a state of balance or coherence among a set of beliefs arrived at by a process of deliberative mutual adjustment among general principles and particular judgments."* In practical terms, reflective equilibrium is about how we identify and resolve logical inconsistencies in our prevailing moral compass. Examples such as the rejection of slavery and of innumerable "isms" (sexism, ageism, etc.) are quite clear: The arguments that worked best were those highlighting the hypocrisy of maintaining acceptance of existing attitudes in the face of already established contrasting attitudes in matters that were indisputably analogous.

Reflective equilibrium gets my vote for the most elegant and beautiful explanation, because of its immense breadth of applicability and also its lack of dependence on other controversial

---

* *A Theory of Justice* (Cambridge, MA: Belknap Press, 1971).

positions. Most important, it rises above the question of cognitivism, the debate over whether there is any such thing as objective morality. Cognitivists assert that certain acts are inherently good or bad, regardless of the society in which they do or do not occur—very much as the laws of physics are generally believed to be independent of those observing their effects. Noncognitivists claim, by contrast, that no moral position is universal and that each (hypothetical) society makes its own moral rules unfettered, so that even acts we would view as unequivocally immoral could be morally unobjectionable in some other culture. But when we make actual decisions concerning whether such-and-such a view is morally acceptable or not, reflective equilibrium frees us from the need to take a view on the cognitivism question. In a nutshell, it explains why we don't need to know whether morality is objective.

I highlight monogamy here because, of the many topics to which reflective equilibrium can be usefully applied, Western society's position on monogamy is at the most critical juncture. Monogamy today compares with heterosexuality not too many decades ago, or tolerance of slavery 150 years ago. Quite a lot of people depart from it, a much smaller minority actively advocate the acceptance of departure from it, but most people advocate it and disparage the minority view. Why is this the "critical juncture"? Because it is the point at which enlightened thought-leaders can make the greatest difference to the speed with which the transition to the morally inescapable position occurs.

First let me make clear that I refer here to sex and not (necessarily, anyway) to deeper emotional attachments. Whatever one's views or predilections concerning the acceptability or desirability of having deep emotional attachments with more than one partner, fulfillment of the responsibilities they entail tends to take a significant proportion of the twenty-four hours of everyone's day. The complications arising from this inconvenient truth are a topic for another time. In this essay, I focus on liaisons casual enough (whether or not repeated) that availability of time is not a major issue.

An argument from reflective equilibrium always begins with identification of the conventional views, with which one then makes a parallel. In this case, it's all about jealousy and possessiveness. Consider chess, or drinking. These are rarely solitary pursuits. Now, is it generally considered reasonable for a friend with whom one sometimes plays chess to feel aggrieved when one plays chess with someone else? Indeed, if someone exhibited possessiveness in such a matter, would they not be viewed as unacceptably overbearing and egotistical?

My claim is probably obvious by now. It is simply that there is nothing about sex that morally distinguishes it from other activities performed by two (or more) people collectively. In a world no longer driven by reproductive efficiency, and presuming that all parties are taking appropriate precautions in relation to pregnancy and disease, sex is overwhelmingly a recreational activity. What, then, can morally distinguish it from other recreational activities? Once we see that nothing does, reflective equilibrium forces us to one of two positions: Either we start to resent the temerity of our regular chess opponents playing others, or we cease to resent the equivalent in sex.

My prediction that monogamy's end is extremely nigh arises from my reference to reproductive efficiency above. Every single society in history has seen a precipitous reduction in fertility following its achievement of a level of prosperity that allowed reasonable levels of female education and emancipation. Monogamy is virtually mandated when a woman spends her entire adult life with young children underfoot, because continuous financial support cannot otherwise be ensured. But when it is customary for those of both sexes to be financially independent, this logic collapses. This is especially so for the increasing proportion of men and women who choose to delay having children until middle age (if then).

I realize that rapid change in a society's moral compass needs more than the removal of influences maintaining the status quo; it also needs an active impetus. What is the impetus in this case? It is simply the pain and suffering that arises when the possessiveness and jealousy inherent in the monogamous mind-set butt heads with the asynchronous shifts of affection and aspiration inherent in the response of human beings to their evolving social interactions. Gratuitous suffering is anathema to all. Thus, the realization that this particular category of suffering is wholly gratuitous has not only irresistible moral force (via the principle of reflective equilibrium) but also immense emotional utility.

The writing is on the wall.

BOLTZMANN'S EXPLANATION OF THE SECOND LAW OF THERMODYNAMICS

LEONARD SUSSKIND

Felix Bloch Professor of Physics, Stanford; director, Stanford Institute for Theoretical Physics; author, The Black Hole War: My Battle with Stephen Hawking to Make the World Safe for Quantum Mechanics

"What is your favorite deep, elegant, or beautiful explanation?" That's a tough question for a theoretical physicist; theoretical physics is all about deep, elegant, beautiful explanations, and there are many to choose from.

Personally, my favorites are explanations that get a lot for a little. In physics, that means a simple equation or a very general principle. I have to admit, though, that no equation or principle appeals to me more than Darwinian evolution, with the selfish-gene mechanism thrown in. To me, it has what the best physics explanations have: a kind of mathematical inevitability. But there are many people who can explain evolution better than I, so I will stick to what I know best.

The guiding star for me, as a physicist, has always been Ludwig Boltzmann's explanation of the second law of thermodynamics—the law that says that entropy never decreases. To the physicists of the late 19th century, this was a very serious paradox. Nature is full of irreversible phenomena—things that easily happen but could not possibly happen in reverse order. However, the fundamental laws of physics are completely reversible: Any solution of Newton's equations can be run backwards and it's still a solution. So if entropy can increase, the laws of physics say it must be able to decrease. But experience says otherwise. For example, if you watch a movie of a nuclear explosion in reverse, you know very well that it's fake. As a rule, things go one way and not the other. Entropy increases.

What Boltzmann realized is that the second law—entropy never decreases—is not a law in the same sense as Newton's law of gravity or Faraday's law of induction. It's a probabilistic law that has the same status as the following obvious claim: If you flip a coin a million times, you will not get a million heads. It simply won't happen. But is it possible? Yes, it is; it violates no law of physics. Is it likely? Not at all. Boltzmann's formulation of the second law was very similar. Instead of saying entropy does not decrease, he said entropy *probably* doesn't decrease. But if you wait around long enough in a closed environment, you will eventually see entropy decrease; by accident, particles and dust will come together and form a perfectly assembled bomb. How long? According to Boltzmann's principles, the answer is the exponential of the entropy created when the bomb explodes. That's a very long time, a lot longer than the time it takes to flip a million heads in a row.

I'll give you a simple example to see how it's possible for things to be more probable one way than the other, despite both being possible. Imagine a high hill that comes to a narrow point—a needle point—at the top. Now imagine a bowling ball balanced at the top of the hill. A tiny breeze comes along. The ball rolls off the hill, and you catch it at the bottom. Next, run it in reverse: The ball leaves your hand, rolls up the hill, and with infinite finesse, comes to the top—and stops! Is it possible? It is. Is it likely? It is not. You would have to have almost perfect precision to get the ball to the top, let alone to have it stop dead-balanced. The same is true with the bomb. If you could reverse every atom and particle with sufficient accuracy, you could make the explosion products reassemble themselves. But a tiny inaccuracy in the motion of just one single particle and all you would get is more junk.

Here's another example: Drop a bit of black ink into a tub of water. The ink spreads out and eventually makes the water gray. Will a tub of gray water ever clear up and produce a small drop of ink? Not impossible, but very unlikely.

Boltzmann was the first to understand the statistical foundation for the second law, but he was also the first to understand the inadequacy of his own formulation. Suppose you came upon a tub that had been filled a zillion years ago and had not been disturbed since. You notice the odd fact that it contains a somewhat localized cloud of ink. The first thing you might ask is, What will happen next? The answer is that the ink will almost certainly spread out more. But by the same token, if you ask what most likely took place a moment before, the answer would be the same: It was probably more spread out a moment ago than it is now. The most likely explanation would be that the ink blob is just a momentary fluctuation.

Actually, I don't think you'd come to that conclusion at all. A much more reasonable explanation is that, for reasons unknown, the tub started not so long ago with a concentrated drop of ink, which then spread. Understanding why ink and water go one way becomes a problem of "initial conditions." What set up the concentration of ink in the first place?

The water and ink is an analogy for the question of why entropy increases. It increases because it's most likely that it will increase. But the equations say that it's also most likely that it increases toward the past. To understand why we have this sense of direction, one must ask the same question Boltzmann did: Why was the entropy very small at the beginning? What created the universe in such a special low-entropy way? That's a cosmological question we are still very uncertain about.

I began telling you what my favorite explanation is, and I ended up telling you what my favorite unsolved problem is. I apologize for not following the instructions. But that's the way of all good explanations. The better they are, the more questions they raise.

THE DARK MATTER OF THE MIND

JOEL GOLD

Psychiatrist; clinical associate professor of psychiatry, NYU School of Medicine

There are people who want a stable marriage yet continue to cheat on their wives.

There are people who want a successful career yet continue to undermine themselves at work.

Aristotle defined man as a rational animal. Contradictions like these show that we are not. All people live with the conflicts between what they want and how they live. For most of human history we had no way to explain this paradox, until Freud's discovery of the unconscious resolved it. Before Freud, we were restricted to our conscious awareness when looking for answers regarding what we knew and felt. All we had to explain incompatible thoughts, feelings, and motivations was limited to what we could access in consciousness. We knew what we knew and we felt what we felt. Freud's elegant explanation postulated a conceptual space, not manifest to us, where irrationality rules. This aspect of the mind is not subject to the constraints of rationality, such as logical inference, cause and effect, and linear time. The unconscious explains why presumably rational people live irrational lives.

Critics may take exception as to what Freud believed resides in the unconscious—drives both sexual and aggressive, defenses, conflicts, fantasies, affects, and beliefs—but no one would deny its existence; the unconscious is now a commonplace. How else to explain our stumbling through life, unsure of our motivations, inscrutable to ourselves? I wonder what a behaviorist believes is at play while he is in the midst of divorcing his third astigmatic redhead.

The universe consists primarily of dark matter. We can't see it, but it has an enormous gravitational force. The conscious mind—much like the visible aspect of the universe—is only a small fraction of the mental world. The dark matter of the mind, the unconscious, has the greatest psychic gravity. Disregard the dark matter of the universe and anomalies appear. Ignore the dark matter of the mind and our irrationality is inexplicable.

"THERE ARE MORE THINGS IN HEAVEN AND EARTH . . . THAN ARE DREAMT OF IN YOUR PHILOSOPHY."

ALAN ALDA

Actor, writer, director; host of PBS program The Human Spark; *author,* Things I Overheard While Talking to Myself

That doesn't sound like an explanation, but I take it that way. For me, Hamlet's admonition explains the confusion and uncertainty of the universe (and, lately, the multiverse). It urges us on when, as they always will, our philosophies produce anomalies. It answers the unspoken question, "WTF?" With every door into nature we nudge open, 100 new doors become visible, each with its own inscrutable combination lock. It is both an explanation and a challenge, because there's always more to know.

I like the way it endlessly loops back on itself. Every time you discover a new thing in heaven or earth, it becomes part of your philosophy, which will eventually be challenged by new new things.

Like all explanations, of course, it has its limits. Hamlet says it to urge Horatio to accept the possibility of ghosts. It could just as well be used to prompt belief in UFOs, astrology, and even God—as if to say that that something is proved to exist by the very fact that you can't disprove it exists.

Still, the phrase can get us places. Not as a taxi to the end of thinking but as a passport to exploration. These words of Hamlet's are best thought of as a *ratchet*—a word earthily beautiful in sound and meaning: Keep moving on, but preserve what works. We need Einstein for GPS, but we can still get to the moon with Newton.

AN UNRESOLVED (AND THEREFORE UNBEAUTIFUL) REACTION TO THE *EDGE* QUESTION

REBECCA NEWBERGER GOLDSTEIN
*Philosopher, novelist; Franke Visiting Fellow, Whitney
Humanities Center, Yale; author,* 36 Arguments for
the Existence of God: A Work of Fiction

This year's *Edge* Question sits uneasily on a deeper question: Where do we get the idea—a fantastic idea, if you stop and think about it—that the beauty of an explanation has anything to do with the likelihood of its being true? What do beauty and truth have to do with each other? Is there any good explanation of why the central notion of aesthetics (fluffy) should be inserted into the central notion of science (rigorous)?

You might think that rather than being a criterion for assessing explanations, the sense of beauty is a phenomenon to be explained *away.* Take, for example, our impression that symmetrical faces and bodies are beautiful. Symmetry, it turns out, is a good indicator of health and consequently of mate-worthiness. It's a significant challenge for an organism to coordinate the production of its billions of cells so that its two sides proceed to develop as perfect matches, warding off disease and escaping injury, mutation, and malnutrition. Symmetrical female breasts, for example, are a good predictor of fertility. As our lustful genes know, the achievement of symmetry is a sign of genetic robustness; we find lopsidedness a turnoff. So, too, in regard to other components of human beauty—radiant skin, shining eyes, neotony (at least in women). The upshot is that we don't want to mate with people because they're beautiful; rather, they're beautiful because we want to mate with them, and we want to mate with them because our genes are betting on them as replicators.

So, too, you might think that beauty of every sort is to be similarly explained away, an attention-grabbing epiphenomenon with no substance of its own. Which brings me to the *Edge* Question concerning beautiful explanations. Is there anything to this notion of explanatory beauty, a guide to choosing between explanatory alternatives, or is it just that any explanation that's satisfactory will, *for that very reason and no other*, strike us as beautiful, beautifully *explanatory*, so that the reference to beauty is, once again, without any substance? That would be an explanation for the mysterious injection of aesthetics into science. The upshot would be that explanations aren't satisfying because they're beautiful; rather, they're beautiful because they're satisfying. They strip the phenomenon bare of all mystery and maybe, as a bonus, pull in further phenomena which can be rendered nonmysterious using the same sort of explanation. Can explanatory beauty be explained away, summarily dismissed by way of an eliminative explanation? (Eliminative explanations are beautiful.)

I'd like to stop here, with a beautiful explanation for explaining away explanatory beauty, but somebody is whispering in my ear. It's that damned Plato. Plato is going on about how there is more in the idea of explanatory beauty than is acknowledged in the eliminative explanation. In particular, he's insisting, as he does in his *Timaeus*, that the beauty of symmetry, especially as it's expressed in the mathematics of physical laws, cannot be explained away with the legerdemain of the preceding paragraph. He's reproaching the eliminative explanation of explanatory beauty with ignoring the many examples in history when the insistence on the beauty of symmetry led to substantive scientific progress. What was it that led James Clerk Maxwell to his four equations of electromagnetism but his trying to impose mathematical symmetry on the domains of electricity and magnetism? What was it that led Einstein to his equations of gravity but an insistence on beautiful mathematics?

Eliminative explanations are beautiful, but only when they truly and thoroughly explain. So instead of offering an answer to this year's *Edge* Question, I offer instead an unresolved (and, therefore, unbeautiful) reaction to the deep question on which it rests.

PTOLEMY'S UNIVERSE

JAMES J. O'DONNELL
Classicist; provost, Georgetown University; author,
The Ruin of the Roman Empire

Claudius Ptolemy explained the sky. He was an Egyptian who wrote in Greek in the Roman Empire, in the time of emperors like Trajan and Hadrian. His most famous book was called by its Arabic translators the *Almagest.* He inherited a long ancient tradition of astronomical science going back to Mesopotamia, but he put his name and imprint on the most successful and so far longest-lived mathematical description of the working of the skies.

Ptolemy's geocentric universe is now known mainly as the thing that was rightly abandoned by Copernicus, Kepler, Newton, and Einstein, in progressive waves of the advancement of modern science, but he deserves our deep admiration. Ptolemy's universe actually made sense. He knew the difference between planets and stars and he knew that the planets take some explaining. (The Greek word *planet* means "wanderer," to reflect ancient puzzlement that those bright lights moved according to no pattern that a shepherd or seaman could intuitively predict, unlikely the reassuringly confident annual march of Orion or the rotation of the great bears overhead.) So Ptolemy represents the heavenly machine in a complex mathematical system most notorious for its "epicycles"—the orbits within orbits, so to speak, by which the planets, while orbiting the Earth, spun off their orbits in smaller circles that explained their seeming forward and backward motion in the night sky.

We should admire Ptolemy for many reasons, but chief among them is this: He did his job seriously and responsibly with the tools he had. Given what he knew, his system was brilliantly conceived, mathematically sound, and a huge advance over what had gone before. His observations were patient and careful and as complete as could be, his mathematical calculations correct. More, his mathematical system was as complicated as it needed to be and at the same time as simple as it could be, given what he had to work with. He was, in short, a real scientist. He set the standard.

It took a long time and there were some long arguments before astronomy could advance beyond what he offered—and that's a sign of his achievement. But when advance was possible, Ptolemy had made it impossible for advance to come through wishful thinking, witch doctors, or fantasy. His successors in the great age of modern astronomy had to play by his rules. They needed to observe more carefully, do their math with exacting care, and propose systems at the poise point of complexity and simplicity. Ptolemy challenged the moderns to outdo him—and so they could and did. We owe him a lot.

QUASI-ELEGANCE

PAUL STEINHARDT
Albert Einstein Professor in Science, Departments of
Physics and Astrophysical Sciences, Princeton University;
coauthor (with Neil Turok), Endless Universe

My first exposure to true elegance in science was through a short semi-popular book entitled *Symmetry*, written by the renowned mathematician Hermann Weyl. I discovered the book in the fourth grade and have returned to reread passages every few years. The book begins with the intuitive aesthetic notion of symmetry for the general reader, drawing interesting examples from art, architecture, biological forms, and ornamental design. By the fourth and final chapter, though, Weyl turns from vagary to precise science as he introduces elements of group theory, the mathematics that transforms symmetry into a powerful tool.

To demonstrate its power, Weyl outlines how group theory can be used to explain the shapes of crystals. Crystals have fascinated humans throughout history because of the beautiful faceted shapes they form. Most rocks contain an amalgam of different minerals, each of which is crystalline but which have grown together or crunched together or weathered to the point that facets are unobservable. Occasionally, though, the same minerals form individual large faceted crystals; that's when we find them most aesthetically appealing. "Aluminum oxide" may not sound like something of value, but add a little chromium and give nature sufficient time, and you have a ruby worthy of kings.

The crystal facets found in nature meet only at certain angles corresponding to one of a small set of symmetries. But why does matter take some shapes and not others? What scientific information do the shapes convey? Weyl explains how these questions can be answered by seemingly unrelated abstract mathematics aimed at answering a different question: What shapes can be used to tessellate a plane or fill space if the shapes are identical, meet edge-to-edge, and leave no spaces?

Squares, rectangles, triangles, parallelograms, and hexagons can do the job. Perhaps you imagine that many other polygons would work as well—but try and you will discover there are no more possibilities. Pentagons, heptagons, octagons, and all other regular polygons cannot fit together without leaving spaces. Weyl's little book describes the mathematics allowing a full classification of possibilities; the final tally is only 17 in two dimensions (the so-called wallpaper patterns) and 230 in three dimensions.

The stunning fact about the list was that it precisely matched the list observed for crystals' shapes found in nature. The inference is that crystalline matter is like a tessellation made of indivisible, identical building blocks that repeat to make the entire solid. Of course, we know today that these building blocks are clusters of atoms or molecules. However, bear in mind that the connection between the mathematics and real crystals was made in the 19th century, when the atomic theory was still in doubt. It is amazing that an abstract study of tiles and building blocks can lead to a keen insight about the fundamental constituents of matter and a classification of all possible arrangements of them. It is a classic example of what physicist Eugene Wigner referred to as the "unreasonable effectiveness of mathematics in the natural sciences."

The story does not end there. With the development of quantum mechanics, group theory and symmetry principles have been used to predict the electronic, magnetic, elastic, and other physical properties of solids. Emulating this triumph, physicists have successfully used symmetry principles to explain the fundamental constituents of nuclei and elementary particles, as well as the forces through which they interact.

As a young student reading Weyl's book, I thought crystallography seemed like the ideal of what one should be aiming for in science: elegant mathematics that provides a complete understanding of all physical possibilities. Ironically, many years later, I played a role in showing that my "ideal" was seriously flawed. In 1984, Dan Shechtman, Ilan Blech, Denis Gratias, and John Cahn reported the discovery of a puzzling man-made alloy of aluminum and manganese with icosahedral symmetry.[*] Icosahedral symmetry, with its six fivefold symmetry axes, is the most famous forbidden crystal symmetry. As luck would have it, Dov Levine (Technion) and I had been developing a hypothetical idea of a new form of solid we dubbed *quasicrystals*, short for "quasiperiodic crystals." (A *quasiperiodic* atomic arrangement means the atomic positions can be described by a sum of oscillatory functions whose frequencies have an irrational ratio.) We were inspired by a two-dimensional tiling invented by Sir Roger Penrose known as the Penrose tiling, comprised of two tiles arranged in a fivefold symmetric pattern. We showed that quasicrystals could exist in three dimensions and were not subject to the rules of crystallography. In fact, they could have any of the symmetries forbidden to crystals. Furthermore, we showed that the diffraction patterns predicted for icosahedral quasicrystals matched the Shechtman *et al.* observations.

Since 1984, quasicrystals with other forbidden symmetries have been synthesized in the laboratory. The 2011 Nobel Prize in chemistry was awarded to Dan Shechtman for his experimental breakthrough that changed our thinking about possible forms of matter. More recently, colleagues and I have found evidence

---

[*]   D. Shechtman et al., "Metallic Phase with Long-Range Orientational Order and No Translational Symmetry," *Phys. Rev. Lett.* 53, 1951–3 (1984).

that quasicrystals may have been among the first minerals to have formed in the solar system.

The crystallography I first encountered in Weyl's book, thought to be complete and immutable, turned out to be woefully incomplete, missing literally an uncountable number of possible symmetries for matter. Perhaps there is a lesson to be learned: While elegance and simplicity are often useful criteria for judging theories, they can sometimes mislead us into thinking we are right when we are actually infinitely wrong.

MATHEMATICAL OBJECT OR NATURAL OBJECT?

SHING-TUNG YAU

Mathematician, Harvard University; coauthor (with Steve Nadis), The Shape of Inner Space

Most scientific facts are based on things we cannot see with the naked eye or hear with our ears or feel with our hands. Many of them are described and guided by mathematical theory. In the end, it becomes difficult to distinguish a mathematical object from objects in nature.

One example is the concept of a sphere. Is the sphere part of nature or is it a mathematical artifact? That is difficult for a mathematician to say. Perhaps the abstract mathematical concept is actually a part of nature. And it is not surprising that this abstract concept actually describes nature quite accurately.

SIMPLICITY

FRANK WILCZEK

Theoretical physicist, MIT; corecipient, 2004 Nobel Prize in Physics; author, The Lightness of Being

We all have an intuitive sense of what "simplicity" means. In science, the word is often used as a term of praise. We expect that simple explanations are more natural, sounder, and more reliable than complicated ones. We abhor epicycles, or long lists of exceptions and special cases. But can we take a crucial step further, to refine our intuitions about simplicity into precise, scientific concepts? Is there a simple core to "simplicity"? Is simplicity something we can quantify and measure?

When I think about big philosophical questions, which I probably do more than is good for me, one of my favorite techniques is to try to frame the question in terms that could make sense to a computer. Usually it's a method of destruction: It forces you to be clear, and once you dissipate the fog, you discover that very little of your big philosophical question remains. Here, however, in coming to grips with the nature of simplicity, the technique proved creative, for it led me straight toward a (simple) profound idea in the mathematical theory of information—the idea of description length. The idea goes by several different names in the scientific literature, including algorithmic entropy and Kolmogorov-Smirnov-Chaitin complexity. Naturally I chose the simplest one.

Description length is actually a measure of complexity, but for our purposes that's just as good, since we can define simplicity as the opposite—or, numerically, the negative—of complexity. To ask a computer how complex something is, we have to present that "something" in a form the computer can deal with—that is, as a data file, a string of 0s and 1s. That's hardly a crippling constraint: We know that data files can represent movies, for example, so we can ask about the simplicity of anything we can present in a movie. Since our movie might be a movie recording scientific observations or experiments, we can ask about the simplicity of a scientific explanation.

Interesting data files might be very big, of course. But big files need not be genuinely complex; for example, a file containing trillions of 0s and nothing else isn't genuinely complex. The idea of description length is, simply, that a file is only as complicated as its simplest description. Or, to put it in terms a computer could relate to, a file is as complicated as the shortest program that can produce it from scratch. This defines a precise, widely applicable, numerical measure of simplicity.

An impressive virtue of this notion of simplicity is that it illumines and connects other attractive, successful ideas. Consider, for example, the method of theoretical physics. In theoretical physics, we try to summarize the results of a vast number of observations and experiments in terms of a few powerful laws. We strive, in other words, to produce the shortest possible program that outputs the world. In that precise sense, theoretical physics is a quest for simplicity.

It's appropriate to add that symmetry, a central feature of the physicist's laws, is a powerful simplicity enabler. For example, if we work with laws that have symmetry under space- and time-translation—in other words, laws that apply uniformly, everywhere and everywhen—then we don't need to spell out new laws for distant parts of the universe or for different historical epochs, and we can keep our world-program short.

Simplicity leads to depth: For a short program to unfold into rich consequences, it must support long chains of logic and calculation, which are the essence of depth.

Simplicity leads to elegance: The shortest programs will contain nothing gratuitous. Every bit will play a role, for otherwise we could expunge it and make the program shorter. And the different parts will have to function together smoothly, in order to make a lot from a little. Few processes are more elegant, I think, than the construction, following the program of DNA, of a baby from a fertilized egg.

Simplicity leads to beauty: For it leads, as we've seen, to symmetry, which is an aspect of beauty. As, for that matter, are depth and elegance.

Thus simplicity, properly understood, explains what it is that makes a good explanation deep, elegant, and beautiful.

SIMPLICITY ITSELF

THOMAS METZINGER

Philosophisches Seminar, Johannes Gutenberg-Universität Mainz; author, The Ego Tunnel

Elegance is more than an aesthetic quality or some sort of ephemeral uplifting we experience in deeper forms of intuitive understanding. Elegance is formal beauty. And formal beauty as a philosophical principle is one of the most dangerous, subversive ideas humanity has discovered: It is the virtue of theoretical simplicity. Its destructive force is greater than Darwin's algorithm or that of any other single scientific explanation, because it shows us what the depth of an explanation *is*.

Elegance as theoretical simplicity comes in many different forms. Everybody knows Occam's razor, the ontological principle of parsimony: *Entities are not to be multiplied beyond necessity*. William of Occam gave us a metaphysical principle for choosing between competing theories. All other things being equal, it is rational to prefer the theory that makes fewer ontological assumptions.

We are to admit no more causes of natural things than such as are both true and sufficient to explain their appearances—Isaac Newton formulated this as the First Rule of Reasoning in Philosophy, in his *Principia Mathematica*. Throw out everything that is explanatorily idle, and then shift the burden of proof to the proponent of a less simple theory. In Albert Einstein's words: *The grand aim of all science . . . is to cover the greatest possible number of empirical facts by logical deductions from the smallest possible number of hypotheses or axioms*.

Of course, in today's technical debates new questions have emerged: Why do metaphysics at all? Isn't what we should measure simply the number of free, adjustable parameters in competing hypotheses? Isn't it syntactic simplicity that captures elegance best in, say, the number-fundamental abstractions and guiding principles a theory makes use of? Or will the true criterion of elegance ultimately be found in statistics—in selecting the best model for a set of data points while optimally balancing parsimony with the "goodness of fit" of a suitable curve? And, of course, for Occam-style ontological simplicity, the big question remains: Why should a parsimonious theory more likely be true? Ultimately, isn't all of this rooted in a deeply hidden belief that God must have created a beautiful universe?

I find it fascinating to see how the idea of simplicity has kept its force over the centuries. As a metatheoretical principle, it has demonstrated great power—the subversive power of reason and reductive explanation. The formal beauty of theoretical simplicity is deadly and creative at the same time. It destroys superfluous assumptions whose falsity we just cannot bring ourselves to believe, whereas truly elegant explanations always give birth to an entirely new way of looking at the world. What I would really like to know is this: Can the fundamental insight—the destructive, creative virtue of simplicity—be transposed from the realm of scientific explanation into culture or onto the level of conscious experience? What kind of formal simplicity would make our culture a deeper, more beautiful culture? And what is an elegant mind?

EINSTEIN EXPLAINS WHY GRAVITY IS UNIVERSAL

SEAN CARROLL
Theoretical physicist, Caltech; author, From Eternity to
Here: The Quest for the Ultimate Theory of Time

The ancient Greeks believed that heavier objects fall faster than lighter ones. They had good reason to do so; a heavy stone falls quickly, while a piece of paper flutters gently to the ground. But a thought experiment by Galileo pointed out a flaw. Imagine taking the piece of paper and tying it to the stone. Together, the new system is heavier than either of its components and should fall faster. But in reality, the piece of paper slows down the descent of the stone.

Galileo argued that the rate at which objects fall would actually be a universal quantity, independent of their mass or their composition, if it weren't for the interference of air resistance. Apollo 15 astronaut Dave Scott illustrated this point by dropping a feather and a hammer while standing in near-vacuum on the surface of the moon; as Galileo predicted, they fell at the same rate.

Many scientists wondered why this should be the case. In contrast to gravity, particles in an electric field respond in various ways; positive charges are pushed one way, negative charges the other, and neutral particles not at all. But gravity is universal; everything responds to it in the same way.

Thinking about this problem led Albert Einstein to what he called "the happiest thought of my life." Imagine an astronaut in a spaceship with no windows or other way to see the outside world. If the ship is far away from any stars or planets, everything inside will be in free fall; there will be no gravitational field to push them around. Now put the ship in orbit around a massive object, where gravity is considerable. Everything inside will still be in free fall, because all objects are affected by gravity in the same way; no one object is pushed toward or away from any other one. Given just what is observed inside the spaceship, there's no way we could detect the existence of gravity.

Einstein, in his genius, realized the profound implication of this situation: If gravity affects everything equally, it's not right to think of gravity as a "force" at all. Rather, gravity is a feature of spacetime itself, through which all objects move. In particular, gravity is the curvature of spacetime. The space and time through which we move are not fixed and absolute, as Newton had it; they bend and stretch because of the influence of matter and energy. In response, objects are pushed in different directions by spacetime's curvature, a phenomenon we call "gravity." Using a combination of intimidating mathematics and unparalleled physical intuition, Einstein was able to explain a puzzle unsolved since Galileo's time.

EVOLUTIONARY GENETICS AND THE CONFLICTS OF HUMAN SOCIAL LIFE

STEVEN PINKER

Johnstone Family Professor, Department of Psychology, Harvard University; author, The Better Angels of Our Nature

Complex life is a product of natural selection, which is driven by competition among replicators. The outcome depends on which replicators best mobilize the energy and materials necessary to copy themselves and on how rapidly they can make copies which in turn can replicate. The first aspect of the competition may be called survival, metabolism, or somatic effort; the second, replication or reproductive effort. Life at every scale, from RNA and DNA to whole organisms, implements features that execute—and constantly trade off—these two functions.

Among life's tradeoffs is whether to allocate resources (energy, food, risk, time) to pumping out as many offspring as possible and letting them fend for themselves or eking out fewer descendants and enhancing the chances of survival and reproduction of each one. The continuum represents the degree of *parental investment* expended by an organism.

Since parental investment is finite, investing organisms face a second tradeoff, between investing resources in a given offspring and conserving those resources to invest in its existing or potential siblings.

Because of the essential difference between the sexes—females produce fewer but more expensive gametes—the females of most species invest more in offspring than do the males, whose investment is often close to zero. Mammalian females in particular have opted for massive investment, starting with internal gestation and lactation. In some species, including *Homo sapiens*, the males may invest, too, though less than the females.

Natural selection favors the allocation of resources not just from parents to offspring but among genetic kin, such as siblings and cousins. Just as a gene that encourages a parent to invest in offspring will be favoring a copy of itself that sits inside those offspring, so a gene that encourages an organism to invest in a brother or cousin will, some proportion of the time, be helping a copy of itself and will be selected in proportion to the benefits conferred, the costs incurred, and the degree of genetic relatedness.

I've just reviewed the fundamental features of life on Earth (and possibly life everywhere), with the barest mention of contingent facts about our own species—only that we're mammals with male parental investment. I'll add a second: that we're a brainy species that deals with life's conundrums not just with fixed adaptations selected over evolutionary time but also with facultative adaptations (cognition, language, socialization) that we deploy in our lifetimes and whose products we share via culture.

From these deep principles about the nature of the evolutionary process, one can deduce a vast amount about the social life of our species. (Credit where it's due: William Hamilton, George Williams, Robert Trivers, Donald Symons, Richard Alexander, Martin Daly, Margo Wilson.)

- Conflict is a part of the human condition. Notwithstanding religious myths of Eden, romantic images of noble savages, utopian dreams of perfect harmony, and gluey metaphors like attachment, bonding, and cohesion, human life is never free of friction. All societies have some degree of differential prestige and status, inequality of power and wealth, punishment, sexual regulations, sexual jealousy, hostility to other groups, and conflict within the group, including violence, rape, and homicide. Our cognitive and moral obsessions track these conflicts. There is a small number of plots in the world's fiction, defined by adversaries (often murderous) and tragedies of kinship or love (or both). In the real world, our life stories are largely stories of conflict: the hurts, guilts, and rivalries inflicted by friends, relatives, and competitors.

- The main refuge from this conflict is the family—collections of individuals with an evolutionary interest in one another's flourishing. Thus we find that traditional societies are organized around kinship and that political leaders, from great emperors to tinpot tyrants, seek to transfer power to their offspring. Extreme forms of altruism, such as donating an organ or making a risky loan, are typically offered to relatives, as are bequests of wealth after death—a major cause of economic inequality. Nepotism constantly threatens social institutions such as religions, governments, and businesses that compete with the instinctive bonds of family.

- Even families are not perfect havens from conflict, because the solidarity from shared genes must contend with competition over parental investment. Parents have to apportion their investment across all their children, born and unborn, with every offspring equally valuable (all else being equal). But while an offspring has an interest in its siblings' welfare, since it shares half its genes with each full sib, it shares *all* of its genes with itself, so it has a disproportionate interest in its own welfare. The implicit conflict plays itself out throughout the life span: in postpartum depression, infanticide, cuteness, weaning, brattiness, tantrums, sibling rivalry, and struggles over inheritance.

- Sex is not entirely a pastime of mutual pleasure between consenting adults. That is because the different minimal parental investment of men and women translates into differences in their ultimate evolutionary interests. Men, but not women, can multiply their reproductive output with multiple partners. Men are more prone than women to infidelity. Women are more vulnerable than men to desertion. Sex therefore takes place in the shadow of exploitation, illegitimacy, jealousy, spousal abuse, cuckoldry, desertion, harassment, and rape.

- Love is not all you need, and does not make the world go round. Marriage does offer the couple the theoretical possibility of a perfect overlap of genetic interest, and hence an opportunity for the bliss we associate with romantic love, because their genetic fates are bound together in the same package—namely, their children. Unfortunately those interests can diverge because of infidelity, stepchildren, in-laws, or age differences—which are, not coincidentally, major sources of marital strife.

None of this implies that people are robots controlled by their genes, that complex traits are determined by single genes, that people may be morally excused for fighting, raping, or philandering, that people should try to have as many babies as possible, or that people are impervious to influences from their culture (to take some of the common misunderstandings of evolutionary explanations). What it does mean is that a large number of recurring forms of human conflict fall out of a small number of features of the process that made life possible.

THE FAURIE-RAYMOND HYPOTHESIS

JONATHAN GOTTSCHALL

Literary scholar; adjunct instructor, English Department, Washington & Jefferson College; author, The Storytelling Animal

I read about the Faurie-Raymond hypothesis a long time ago, but it didn't click with me until I fought big Nick. Nick is a national guardsman who trains with me at the local mixed martial-arts academy. Technically we were just sparring, not fighting. But Nick is so strong, his punches so sincere, that even when he tries to throw gentle, he makes your consciousness wobble. The bell rang, and we engaged, and my fear passed quickly into disorientation. Something wasn't right. Nick is powerful, but he's not more skillful than I am and he's not what you would call a graceful mover or a sophisticated striker. Nick plows forward: jab, cross; jab, cross, hook. Nick doesn't bob. Nick doesn't weave. Nick plows forward.

So why couldn't I hit him? Why were my punches grazing harmlessly past his temples or glancing off his belly? And why, whenever I tried to slip and counter, was I eating glove leather? I tracked him through the blur of his hands, and all of the angles looked wrong, the planes of his face and body askew. There was nothing solid to hit. And all the while he was hammering me with punches I sensed too late—slow and heavy blows, maddeningly oblique.

When the bell finally saved me, we embraced (it's a paradox; nothing makes men love each other as much as a good-natured fistfight). I collapsed in one of the folding chairs with my head throbbing and the sweat rolling down, and I said to myself, "That seals it. Faurie-Raymond has to be true."

Nick represents a type that 90 percent of boxers fear and despise on sight. Nick is a lefty, which is, according to my pugilism professor, "an abomination" and "a birth defect." Here, my professor joins other righty authorities in the sweet science, who don't seem to be kidding when they say, "All southpaws should be drowned at birth."

My professor's claim that lefties are defective has a surprising grain of truth. In a world of scissors and schooldesks shaped for righties, being a lefty is not just annoying, it seems to be bad for you. According to a number of studies, lefties are at higher risk for disorders like schizophrenia, mental retardation, immune deficiency, epilepsy, learning disability, spinal deformity, hypertension, ADHD, alcoholism, and stuttering.

Which brings me to Charlotte Faurie and Michel Raymond, a pair of French scientists who study the evolution of handedness. Left-handedness is partly heritable and is associated with significant health risks. So why, they wondered, hadn't natural selection trimmed it away? Were the costs of left-handedness canceled by hidden fitness benefits?

They noted that lefties have advantages in sports like baseball and fencing, where the competition is interactive, but not in sports like gymnastics or swimming, with no direct interaction. In the elite ranks of cricket, boxing, wrestling, tennis, baseball, and more, lefties are massively overrepresented. The reason is obvious: Since 90 percent of the world is right-handed, righties usually compete against one another. When they confront lefties, who do everything backwards, their brains reel, and the result can be as lopsided as my mauling by Nick. In contrast, lefties are used to facing righties; when two lefties face off, any confusion cancels out.

Faurie and Raymond made a mental leap. The lives of ancestral people were typically more violent than our own. Wouldn't the lefty advantage in sports—including combat sports like boxing, wrestling, and fencing—have extended to fighting, whether with fists, clubs, or spears? Could the fitness benefits of fighting southpaw have offset the health costs associated with left-handedness? In 2005, they published a paper supporting their prediction of a strong correlation between violence and handedness in preindustrial societies: The more violent the society, the more lefties. The most violent society they sampled, the Eipo of highland New Guinea, was almost 30 percent southpaw.*

What makes a scientific explanation beautiful? General factors like parsimony play a role, but as with any aesthetic question, quirks of personal taste bulk large. Why do I find the Faurie-Raymond hypothesis attractive? Partly because it was an almost recklessly creative idea, and yet the data seemed to fit. But mainly because the undoubtable truth of it was pounded into my brain by a young soldier sometime last year.

This is not to say, with apologies to Keats, that beauty and truth are synonyms. Sometimes the truth turns out to be dull and flat. Many of the loveliest explanations—the ones we adore with almost parental fondness—turn out to be dead false. This is what T. H. Huxley called scientific tragedy, "the slaying of a beautiful hypothesis by an ugly fact." Many studies have since examined the Faurie-Raymond hypothesis. Results have been mixed, but facts have surfaced that are, to my taste, quite decidedly ugly. A recent and impressive inquiry found no evidence that lefties are overrepresented among the Eipo of highland New Guinea.†

It hurts to surrender a beloved idea, one you just *knew* was true, one that was stamped into your mind by lived experience, not statistics. And I'm not yet ready to consign this one to the bone-

---

* Charlotte Faurie & Michel Raymond, "Handedness, homicide and negative frequency-dependent selection," *Proc. Roy. Soc. B* 272, 25-8 (2005).

† Sara M. Schaafsma et al., "Handedness in a nonindustrial society challenges the fighting hypothesis as an evolutionary explanation for left-handedness," *Evol. & Hum. Behavior* 33:2, 94-9 (2012).

yard of lovely but dead science. Faurie and Raymond brought in sports data to shore up their main story about fighting. But I think the sports data may actually be the main story. Lefty genes may have survived more through southpaw success in *play* fights than in real fights—a possibility Faurie and Raymond acknowledge in a later paper.* Athletic contests are important across cultures. Around the world, sport is mainly a male preserve, and winners— from captains of football teams to traditional African wrestlers to Native American runners and lacrosse players—gain more than mere laurels: They elevate their cultural status; they win the admiration of men and the desire of women (research confirms the stereotype: Athletic men have more sexual success). This raises a broader possibility—that our species has been shaped more than we know by the survival of the sportiest.

---

* V. Llaurens, M. Raymond & C. Faurie, "Why Are Some People Left-Handed? An Evolutionary Perspective," *Phil. Trans. Roy. Soc. B* 364, 881-94 (2009).

GROUP POLARIZATION

DAVID G. MYERS

Professor of psychology, Hope College; author, Psychology, *10th edition*

Forty-five years ago, some social-psychological experiments posed story problems that assessed people's willingness to take risks: For example, what odds of success should a budding writer have in order to forgo her sure income and attempt writing a significant novel? To everyone's amazement, group discussions led people to opt for more risk, setting off a wave of speculation about group risk-taking by juries, business boards, and the military. Alas, some other story problems surfaced in which group deliberation increased caution. (Should a young married parent with two children gamble his savings on a hot stock tip?) Out of this befuddlement—Does group interaction increase risk or caution?—there emerged a deeper principle of simple elegance: *Group interaction tends to amplify people's initial inclinations.* This group polarization phenomenon was repeatedly confirmed. In one study, relatively prejudiced and unprejudiced students were grouped separately and asked to respond—before and after discussion—to racial dilemmas, such as a conflict over property rights versus open housing. Discussion with like-minded peers increased the attitude gap between the high- and low-prejudiced groups.

Today, the self-segregation of kindred spirits is rife. With increased mobility, conservative communities attract conservatives and progressive communities attract progressives. Political journalist Bill Bishop and sociologist Robert Cushing report that the percentage of landslide counties—those voting 60 percent or more for one presidential candidate—nearly doubled between

1976 and 2008.* And when neighborhoods become political echo chambers, the consequence is increased polarization, as David Schkade of University of California–San Diego and colleagues demonstrated by assembling small groups of Coloradoans in liberal Boulder and conservative Colorado Springs. The community discussions of climate change, affirmative action, and same-sex unions diverged Boulder folks further leftward and Colorado Springs folks further rightward.

Terrorism is group polarization writ large. Virtually never does it erupt suddenly, as a solo personal act. Rather, terrorist impulses arise among people whose shared grievances bring them together. In isolation from moderating influences, group interaction becomes a social amplifier. The Internet accelerates opportunities for like-minded peacemakers and neo-Nazis, geeks and goths, conspiracy schemers, and cancer survivors to find and influence one another. When socially networked, birds of a feather find their shared interests, attitudes, and suspicions magnified.

Ergo, one elegant and socially significant explanation of diverse observations is simply this: opinion-segregation + conversation → polarization.

* Bill Bishop & Robert G. Cushing, *The Big Sort: Why the Clustering of Like-Minded America Is Tearing Us Apart* (New York: Houghton Mifflin, 2008).

THE PRICE EQUATION

ARMAND MARIE LEROI

Professor of evolutionary developmental biology, Imperial College, London; author, Mutants: On Genetic Variety and the Human Body

Whenever we see highly ordered phenomena—a baby, a symphony, a scientific paper, a corporation, a government, a galaxy—we are driven to ask: How does that order arise? One answer, albeit an abstract one, is that each of these is the product of a variation-selection process. By this I mean any process that begins with many variants and in which most die (or are thrown into the wastepaper basket or dissipate or collapse), leaving only a few fit (or strong or appealing or stable) enough to survive. The production of organic forms by natural selection is, of course, the most famous example of such a process. It's also now a commonplace that human culture is driven by an analogous process; but, as the above examples suggest, variation-selection processes can be seen everywhere, once we know what to look for.

Many others have had this idea, but none has seen its implications as deeply as George Price, an American living in London, who in 1970 published an equation describing variation-selection processes of all kinds.* The Price equation, as it is now known, is simple, deep, and elegant—my candidate explanation. It can be used to describe, *inter alia*, the tuning of an analog radio dial, chemical-reaction kinetics, the impact of neonatal mortality on the distribution of human birth weight, the reason we inhabit *this* universe out of the multitude we do not (assuming the others exist). But for me the real fascination of the Price equation lies not

---

* "Selection and covariance," *Nature* 227, 520-21 (1970).

in the form he gave it in 1970 but in an extension he published two years later.[*]

One of the properties of variation-selection systems is that the selecting can happen at many different levels. Music is clearly the result of a variation-selection process. The composer sits at his piano considering what comes next and chooses one out of the world of possible notes, chords, or phrases that he might. Look at Beethoven's manuscripts (Op. 47, the Kreutzer Sonata, is a good example)—they're scrawled with his second thoughts. In 1996, Brian Eno wittily made this process explicit when he used SSEYO's Koan software to produce an ever-varying collection of pieces that he called "generative music."

But the music we have on our iPods is, of course, not merely the result of the composer's selective choices—nor even those made by producers, performers, and so on—but ours. As individual consumers, we, too, are a selective and hence creative force. And we do not act only as individuals but also as members of social groups. Experiments show that if we know what music other people are listening to, we are quite ready to subsume (if not totally abandon) our own aesthetic preferences and follow the herd—a phenomenon that explains why it's so hard to predict hits. So composers, consumers, and groups of consumers all shape the world of music. Umberto Eco made much the same point as long ago as 1962, in *Opera Aperta* (*The Open Work*). Of course, as a literary critic Eco could do no more than draw attention to the problem. But George Price solved it.

In 1972, Price extended his general variation-selection equation to allow for multilevel selection. This form of the equation has been useful to evolutionary biologists, allowing them to see, for

---

* "Extension of covariance selection mathematics," *Ann. Hum. Genetics* 35:4, 485–90 (1972).

example, the relationship between kin- and group-selection clearly and so put to rest endless controversies stemming from incompatible mathematical formulations. It hasn't yet been applied to cultural evolution, though it surely will be. But the extended Price equation is much more important than even that. It slices one of the Gordian knots that scientists and philosophers of science have long wrestled with.

This is the knot of reducibility. Can the behavior of a system be understood in terms of—that is, be reduced to—the behavior of its components? This question, in one form or another, pervades science. Systems biologists vs. biochemists, cognitive scientists vs. neuroscientists, Durkheim vs. Bentham, Gould vs. Dawkins, Aristotle vs. Democritus—the gulf (epistemological, ontological, and methodological) between the holist vs. reductionist stances lies at the root of many of science's greatest disputes. It is also the source of advances, as one stance is abandoned in favor of another. Indeed, holist and reductionist research programs often exist side by side in uneasy truce (think of any biology department). But when, as so often, the truce breaks down and open warfare resumes, it's clear that what's needed is a way of rationally partitioning the creative forces operating at different levels.

That is what Price gave. His equation applies only to variation-selection systems, but if you think about it, most order-creating systems *are* variation-selection systems. Returning to our musical world: Who really shapes it? Beethoven's epigones tweaking their MIDI files? Adolescents downloading in the solitude of their bedrooms? The massed impulses of the public? I think that Price's equation can explain. It certainly has some explaining to do.

UNCONSCIOUS INFERENCES

GERD GIGERENZER

Psychologist; director of the Center for Adaptive Behavior and Cognition at the Max Planck Institute for Human Development, Berlin; author, Gut Feelings: The Intelligence of the Unconscious

Optical illusions are a pleasure to look at, puzzling, and robust. Even if you know better, you still are caught in the illusion. Why do they exist? Are they merely mental quirks? The physicist and physiologist Hermann von Helmholtz (1821–1894) provided us with a beautiful explanation of the nature of perception and how it generates perceptual illusions of depth, space, and other properties. Perception requires smart bets called *unconscious inferences*.

In Volume III of his *Physiological Optics*, Helmholtz recounts a childhood experience:

> I can recall when I was a boy going past the garrison chapel in
> Potsdam, where some people were standing in the belfry. I mistook
> them for dolls and asked my mother to reach up and get them for me,
> which I thought she could do. The circumstances were impressed
> on my memory, because it was by this mistake that I learned to
> understand the law of foreshortening in perspective.

This childhood experience taught Helmholtz that information available from the retina and other sensory organs is not sufficient to reconstruct the world. Size, distance, and other properties need to be inferred from uncertain cues, which in turn have to be learned by experience. Based on this experience, the brain draws unconscious inferences about what a sensation means. In other words, perception is a kind of bet about what's really out there.

But how exactly does this inference work? Helmholtz drew an analogy with probabilistic syllogisms. The major premise is a collection of experiences that are long out of consciousness; the minor premise is the present sensory impression. Consider the "dots illusion" of V. S. Ramachandran and colleagues at the Center for Brain and Cognition, University of California–San Diego:

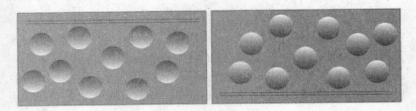

The dots in the left picture appear concave, receding into the surface away from the observer, while those on the right side appear convex, curved toward the observer. If you turn the page around, the inward dots will pop out and vice versa. In fact, the two pictures are identical, except for being rotated 180 degrees. The illusion of concave and convex dots occurs because our brain makes unconscious inferences.

Major premise:
A shade on the upper part of a dot is nearly always associated with a concave shape.

Minor premise:
The shade is in the upper part.

Unconscious inference:
The shape of the dot is concave.

Our brains assume a three-dimensional world, and the major premise guesses the third dimension from two ecological structures:

1. Light comes from above, and

2. There is only one source of light.

These two structures dominated most of human and mammalian history, in which the sun and the moon were the only sources of light, and the first also holds approximately for artificial light today. Helmholtz would have favored the view that the major premise is learned from individual experience; others have favored evolutionary learning. In both cases, visual illusions are seen as the product of unconscious inferences based on evidence that is usually reliable but can be misleading in special circumstances.

The concept of unconscious inference can also explain phenomena from other sensory modalities. A remarkable instance where a major premise suddenly becomes incorrect is the case of a person whose leg has been amputated. Although the major premise ("A stimulation of certain nerves is associated with that toe") no longer holds, patients nevertheless feel pain in toes that are no longer there. The "phantom limb" also illustrates our inability to correct unconscious inferences despite our knowledge. Helmholtz's concept has given us a new perspective on perception in particular and cognition in general:

1. Cognition is inductive inference. Today, the probabilistic syllogism has been replaced by statistical and heuristic models of inference, inspired by Thomas Bayes and Herbert Simon, respectively.

2. Rational inferences need not be conscious. Gut feelings and intuition work with the same inductive inferences as conscious intelligence.

3. Illusions are a necessary consequence of intelligence. Cognition requires going beyond the information given, to make bets and therefore to risk errors. Would we be better off without visual illusions? We would in fact be worse off—like a person who never says anything to avoid making any mistakes. A system that makes no errors is not intelligent.

SNOWFLAKES AND THE MULTIVERSE

MARTIN J. REES

Former president of the Royal Society; emeritus professor of cosmology and astrophysics, University of Cambridge; master, Trinity College; author, From Here to Infinity: A Vision for the Future of Science

An astonishing concept has entered the mainstream of cosmological thought: Physical reality could be hugely more extensive than the patch of space and time traditionally called "the universe." A further Copernican demotion may loom ahead. We've learned that we live in just one planetary system among billions, in one galaxy among billions. But now that's not all. The entire panorama that astronomers can observe could be a tiny part of the aftermath of "our" Big Bang, which is itself just one bang among a perhaps infinite ensemble.

Our cosmic environment could be richly textured but on scales so vast that our purview is restricted to a tiny fragment. We're not aware of the "big picture," any more than a plankton whose universe was a liter of water would be aware of the world's topography and biosphere. It is obviously sensible for cosmologists to start off by exploring the simplest models. But there is no more reason to expect simplicity on the grandest scale than in the terrestrial environment, where intricate complexity prevails.

Moreover, string theorists suspect—for reasons quite independent of cosmology—that there may be an immense variety of "vacuum states." Were this correct, different universes could be governed by different physics. Some of what we call laws of nature may, in this grander perspective, be local bylaws, consistent with some overarching theory governing the ensemble, but not uniquely fixed by that theory. More specifically, some aspects may be arbitrary and others not. As an analogy (which I owe to the astrobiologist and cosmologist Paul Davies), consider the form of snowflakes. Their ubiquitous sixfold symmetry is a direct consequence of the properties and shape of water molecules. But snowflakes display an immense variety of patterns, because each is molded by its distinctive history and microenvironment; how each flake grows is sensitive to the fortuitous temperature and humidity changes during its growth.

If physicists achieved a fundamental theory, it would tell us which aspects of nature are direct consequences of the bedrock theory (just as the symmetrical template of snowflakes is due to the basic structure of a water molecule) and which cosmic numbers are (like the distinctive pattern of a particular snowflake) the outcome of environmental contingencies.

Our domain wouldn't then be just a random one. It would belong to the unusual subset where there was a "lucky draw" of cosmic numbers conducive to the emergence of complexity and consciousness. Its seemingly designed or fine-tuned features wouldn't be surprising. We may, by the end of this century, be able to say with confidence whether we live in a multiverse and how much variety its constituent universes display. The answer to this question will, I think, determine crucially how we should interpret the "biofriendly" universe in which we live (and which we share with any aliens whom we might one day contact).

It may disappoint some physicists if some of the key numbers they're trying to explain turn out to be mere environmental contingencies, no more "fundamental" than the parameters of Earth's orbit around the sun. But that disappointment would surely be outweighed by the revelation that physical reality was grander and richer than hitherto envisioned.

EINSTEIN'S PHOTONS

ANTON ZEILINGER

Physicist, University of Vienna; scientific director, Institute of Quantum Optics and Quantum Information, Austrian Academy of Sciences; author, Dance of the Photons: From Einstein to Quantum Teleportation

My favorite deep, elegant, and beautiful explanation is Albert Einstein's 1905 proposal that light consists of energy quanta, today called photons. It is little known, even among physicists, but extremely interesting how Einstein came to this conclusion. It's often thought that he invented the concept to explain the photoelectric effect. Certainly that is part of Einstein's 1905 publication, but only toward its end. The idea itself is much deeper, more elegant—and, yes, more beautiful.

Imagine a closed container whose walls are at some temperature. The walls are glowing, and as they emit radiation, they also absorb radiation. After some time, there will be an equilibrium distribution of radiation inside the container. This was already well known before Einstein. Max Planck had introduced the idea of quantization that explained the energy distribution of the radiation inside such a volume. Einstein went a step further. He studied how orderly the distribution of the radiation is inside such a container.

For physicists, entropy is a measure of disorder. And the Austrian physicist Ludwig Boltzmann showed that the entropy of a system is a measure of how probable its state is. To take a simple example, it is much more probable that books, notes, pencils, photos, pens, etc. are scattered all over my desk than that they form orderly stacks. Or if we consider a million atoms inside a container, it is much more probable that they are more or less equally distributed throughout the container's volume than all collected in one corner. In both cases, the first state is the less orderly, and when the atoms fill a larger volume they have an even higher entropy.

Einstein realized that the entropy of radiation (including light) changes with the volume of its container in the same mathematical way as for atoms; in both cases, the entropy increases with the logarithm of the volume. For Einstein, this could not be just a coincidence. Since we can understand the entropy of the gas because it consists of atoms, radiation, too, consists of particles—which he called energy quanta (today called photons).

Einstein immediately applied his idea to the photoelectric effect. But he also realized a fundamental conflict of the idea of energy quanta with the well-studied and observed phenomenon of interference.

The problem was how to understand the two-slit interference pattern. This is the phenomenon that, according to Richard Feynman, contains "the only mystery" of quantum physics. The challenge is very simple. When we shine a beam of photons at a plate in which there are two slits, and both slits are open, we obtain bright and dark stripes on an observation screen behind the plate; these are the interference fringes. When we have only one slit open, we get no stripes, no interference fringes, but instead a broad distribution of photons. This result can easily be understood, given the wave picture of light: Waves pass through each of the two slits, alternately extinguishing and reinforcing each other on the observation screen. That way, we obtain dark and bright fringes.

But what to expect if the light beam's intensity is so low that only one photon at a time passes through the apparatus? Following Einstein's realist position, it would be natural to assume that a photon has to pass through either one open slit or the other, but not both. We can do the experiment by sending photons in, one at a time. According to Einstein, no interference fringes should appear, because a single photon, as a particle, would have to "choose" one open slit or the other, and thus there would be no reinforcement or extinguishing, as there was in the wave picture. This was indeed Einstein's opinion, and he suggested that the fringes appear only if many photons are passing through at the same time and somehow interact with each other such that they make up the interference pattern.

Today, we know from many experiments that the interference pattern arises even at such low intensities that only one photon per second passes through the apparatus. If we wait long enough and look at the distribution of all of them on the observation screen, we get the interference pattern. The modern explanation is that the interference pattern arises only if there is no information, anywhere in the universe, about which slit the particle passes through (the colloquial statement that a photon passes through both slits at once has to be taken with a grain of salt). But even as Einstein was wrong here, his idea of the energy quanta of light, i.e., photons, pointed far into the future.

In a letter to his friend Conrad Habicht in the same year of 1905, the miraculous year wherein he also published his special theory of relativity, he called the paper on photons "revolutionary." As far as is known, this was the only work of his that he ever called revolutionary, and therefore it is quite fitting that in 1921 it brought him a Nobel Prize. That the situation was not as clear a few years earlier is witnessed by a famous letter signed by Planck, Walther Nernst, Heinrich Rubens, and Emil Warburg, suggesting Einstein for membership in the Prussian Academy of Sciences in 1913. They wrote: "That he might have in his speculations, occasionally, overshot the target, as for example in his light-quantum hypothesis, should not be counted against him too much, because without occasionally taking a risk, even in the most exact science no real innovation can be introduced." Einstein's deep, elegant, and beautiful explanation in 1905 of the entropy of radiation by proposing light quanta makes a strong case for the usefulness of occasional speculation.

GO SMALL

JEREMY BERNSTEIN

Professor of physics, emeritus, Stevens Institute of Technology; former staff writer, The New Yorker; *author,* Quantum Leaps

When confronted with a question like this, the temptation is to "go big" and respond with something, say, from Einstein's theory of relativity. Instead I will go small. When Planck introduced his quantum of action at the turn of the 20th century, he realized that this allowed for a new set of natural units. For example, the Planck time is the square root of Planck's constant times the gravitational constant divided by the fifth power of the speed of light. It is the smallest unit of time anyone talks about, but is it a "time"? The problem is that these constants are just that. They are the same to a resting observer as to a moving one. But the time is not. I posed this as a "divinette" to my "coven," and Freeman Dyson came up with a beautiful answer. He tried to construct a clock that would measure it. Using the quantum uncertainties, he showed that it would be consumed by a black hole of its own making. No measurement is possible. The Planck time ain't a time—or it may be beyond time.

WHY IS OUR WORLD COMPREHENSIBLE?

ANDREI LINDE

Father of eternal chaotic inflation; professor of physics, Stanford University

"The most incomprehensible thing about the world is that it is comprehensible." So said Albert Einstein. A similar problem was noted by Eugene Wigner, who said that the unreasonable efficiency of mathematics is "a wonderful gift which we neither understand nor deserve."

Why do we live in a comprehensible universe, with certain rules that can be efficiently used for predicting our future?

Of course, one could always respond that this is "just so"—that God created the universe and made it simple enough so that we could comprehend it. But shall we give up so easily? Let us consider several other questions of a similar type. Why is our universe so large? Why don't parallel lines intersect? Why do different parts of the universe look so similar? For a long time, such questions seemed too metaphysical to be considered seriously. Now we know that inflationary cosmology provides a possible answer to all of them.

To understand the issue, let's consider some examples of an incomprehensible universe, where mathematics is inefficient. Suppose the universe is in a state of the so-called Planck density: r ~ 10^{94} g/cm³, which is 94 orders of magnitude greater than the density of water. According to the theory of quantum gravity, quantum fluctuations of spacetime in this regime are so large that all measuring sticks are rapidly bending, shrinking, and extending in a chaotic and unpredictable way—faster than you could measure distance with them. All clocks are destroyed faster than you could measure time with them. All records of previous events become erased, so that you cannot remember anything, record it, and predict the future. The universe is incomprehensible to anybody living there (if life is possible there at all), and the laws of mathematics cannot be efficiently used.

If the huge-density example looks a bit extreme, rest assured that it is not. There are three basic types of universes: closed, open, and flat. A typical closed universe created in the hot Big Bang would collapse in about 10^{-43} seconds into a state with the Planck density, unless it had a huge size to start with. A typical open universe created in the Big Bang would grow so fast that formation of galaxies would be impossible, and our bodies (if we were lucky enough to be born) would be instantly torn apart. Nobody could live in, let alone comprehend, the universe in either of these two cases. We can enjoy life in a flat, or nearly flat, universe (which is what we do now), but unless something special (inflation, see below) happens, this requires fine-tuning of initial conditions at the moment of the Big Bang with an incredible accuracy of about 10^{-60}.

Recent developments in string theory, the most popular candidate for the role of the Theory of Everything, reveal an even broader spectrum of possible but incomprehensible universes. If we assume that our universe is described by string theory, does it mean that we know everything about the world around us? Consider a much simpler example: Recall that water can be liquid, frozen, or gaseous. Chemically it's the same substance, but dolphins can live and comprehend the universe in their own way only if they are surrounded by liquid water. In this example, we have only three choices: liquid, ice, or vapor.

With string theory, according to its latest developments, we may have about 10^{500} (or more) choices of the possible state of the world surrounding us. All of these choices follow from the same basic theory. However, the universes corresponding to each of these choices would look as if they were governed by different laws of physics; their common roots would be well hidden. Since there are so many different choices, some of them, one hopes, can describe the universe we live in. But most of them would lead to a universe in which we could not live, build measuring devices, record events, or efficiently use mathematics and physics to predict the future.

When Einstein and Wigner were trying to understand why our universe is comprehensible and why mathematics is so efficient, everybody assumed that the universe was unique and uniform, and that the laws of physics were the same everywhere. This assumption was called the cosmological principle. We did not know *why* the universe was the same everywhere, we just took it for granted. Thus the problem described by Einstein and Wigner was supposed to apply to the whole universe. In this context, recent developments would only sharpen the formulation of the problem: If a typical universe is hostile to life as we know it, then we must be incredibly lucky to, by chance, live in the universe where life is possible and the universe comprehensible. This would indeed look like a miracle, like a "gift which we neither understand nor deserve." Can we do better than rely on the miraculous?

In the last thirty years, the way we think about the origin and the global structure of our world has changed profoundly. First of all, we found that inflation, the exponentially rapid expansion of the early universe, makes the universe flat and thus potentially suitable for life. Moreover, the rapid stretching of the universe makes the part where we live extremely homogeneous. Thus we have found an explanation for the observed uniformity of the universe. However, we have also found that on a very, very large scale (well beyond the present observable horizon of about 10^{10} light-years), the universe becomes 100 percent nonuniform, because of quantum effects amplified by the explosive expansion of space.

In the context of string theory in combination with inflationary cosmology, this means that instead of looking like an expanding symmetric sphere, our world looks more like a multiverse—an incredibly large collection of exponentially large bubbles. Each one of these bubbles looks like a universe, and now we use the word "universe" to describe enormous, locally uniform parts of the world. One of the 10^{500} different laws of the low-energy physics originating from string theory operates inside each of these universes.

In some of these universes, quantum fluctuations are so large that any computations are impossible; mathematics there is inefficient, because predictions cannot be memorized and used. The lifetimes of some universes are too short. Other universes are long-lived but empty; their laws of physics do not allow the existence of any entities who could survive long enough to learn physics and mathematics.

Fortunately, among all possible parts of the multiverse, there should be some universes where life as we know it is possible. But our life is possible only if the laws of physics operating in our part of the multiverse allow formation of stable, long-lived structures capable of making computations. This implies existence of mathematical relations that can be used for the long-term predictions. The rapid development of the human race was possible only because we live in the part of the multiverse where the long-term predictions are so useful and efficient that they allow us to survive in the hostile environment and win in the competition with other species.

To summarize (and generalize), the inflationary multiverse consists of myriads of "universes" with all possible laws of physics and mathematics operating in each. We can live only in those universes where the laws of physics allow our existence, which requires making reliable predictions. In other words, mathematicians and physicists can live only in those universes that are comprehensible and where the laws of mathematics are efficient.

You can dismiss everything I just wrote as wild speculation. It's interesting, however, that in the context of the new cosmological paradigm developed in the last thirty years we might be able, for the first time, to approach one of the most complicated and mysterious questions that bothered two of the greatest scientists of the 20th century.

ALFVÉN'S COSMOS

GEORGE DYSON

Science historian; author, Turing's Cathedral:
The Origins of the Digital Universe

A hierarchical universe can have an average density of zero, while containing infinite mass.

Hannes Alfvén (1908–1995), who pioneered the field of magnetohydrodynamics, against initial skepticism, to give us a universe permeated by what are now called Alfvén waves, never relinquished his own skepticism concerning the Big Bang. "They fight *against* popular creationism, but at the same time they fight fanatically *for* their own creationism," he argued in 1984,[*] advocating, instead, for a hierarchical cosmology, whose mathematical characterization he credited to Edmund Edward Fournier d'Albe (1868–1933) and Carl Vilhelm Ludvig Charlier (1861–1934). Hierarchical does not mean isotropic, and observed anisotropy does not rule it out.

Gottfried Wilhelm Leibniz (1646–1716), a lawyer as well as a scientist, believed that our universe was selected, out of an infinity of possible universes, to produce maximum diversity from a minimal set of natural laws. It's hard to imagine a more beautiful set of boundary conditions than zero density and infinite mass. But this same principle of maximum diversity warns us that it may take all the time in the universe to work the details out.

---

[*] "Cosmology: Myth or Science?," *Jour. Astrophys. & Astron.* 5, 79-98 (1984).

OUR UNIVERSE GREW LIKE A BABY

MAX TEGMARK

*Cosmologist; associate professor of physics, MIT; scientific
director, Foundational Questions Institute*

What caused our Big Bang? My favorite deep explanation is that
our baby universe grew like a baby human—literally. Right after
your conception, each of your cells doubled roughly daily, causing
your total number of cells to increase day by day as 1, 2, 4, 8, 16,
etc. Repeated doubling is a powerful process, so your mom would
have been in trouble if you'd kept doubling your weight every day
until you were born: After nine months (about 274 doublings), you
would have weighed more than all the matter in our observable
universe combined.

Crazy as it sounds, this is exactly what our baby universe did,
according to the inflation theory pioneered by Alan Guth and
others. Starting out with a speck much smaller and lighter than
an atom, it repeatedly doubled its size until it was more massive
than our entire observable universe, expanding at dizzying speed.
And it doubled not daily but almost instantly. In other words,
inflation created our mighty Big Bang out of almost nothing, in
a tiny fraction of a second. By the time you reached about 10
centimeters in size, your expansion had transitioned from accelerating to decelerating. In the simplest inflation models, our baby
universe did the same when *it* was about 10 centimeters in size,
its exponential growth spurt slowing to a more leisurely expansion wherein hot plasma diluted and cooled and its constituent
particles gradually coalesced into nuclei, atoms, molecules, stars,
and galaxies.

Inflation is like a great magic show. My gut reaction is "This can't possibly obey the laws of physics." Yet under close enough scrutiny, it does. For example, how can one gram of inflating matter turn into two grams when it expands? Surely, mass can't just be created from nothing? Interestingly, Einstein offered us a loophole via his special relativity theory, which says that energy e and mass m are related according to the famous formula $e = mc^2$, where c is the speed of light. This means you can increase the mass of something by adding energy to it. For example, you can make a rubber band slightly heavier by stretching it: You need to apply energy to stretch it, and this energy goes into the rubber band and increases its mass. A rubber band has *negative pressure*, because you need to do work to expand it. Similarly, the inflating substance has to have negative pressure in order to obey the laws of physics, and this negative pressure has to be so huge that the energy required to expand it to twice its volume is exactly enough to double its mass. Remarkably, Einstein's general theory of relativity says that this negative pressure causes a negative gravitational force. This in turn causes the repeated doubling, ultimately creating everything we can observe from almost nothing.

To me, the hallmark of a deep explanation is that it answers more than you ask. And inflation has proven to be the gift that keeps on giving, churning out answer after answer. It explained why space is so flat, which we've measured to about 1 percent accuracy. It explained why, on average, our distant universe looks the same in all directions, with only 0.002 percent fluctuations from place to place. It explained the origins of these 0.002 percent fluctuations as quantum fluctuations stretched by inflation from microscopic to macroscopic scales, then amplified by gravity into today's galaxies and cosmic large-scale structure. It even explained the cosmic acceleration that nabbed the 2011 physics Nobel Prize as inflation, restarting in slow motion, doubling the size of our universe not every split second but every 8 billion years—which has transformed the debate from whether inflation happened or not to whether it happened once or twice.

It's now becoming clear that inflation is an explanation that doesn't stop—inflating or explaining.

Just as cell division didn't make just one baby and stop, but a huge and diverse population of humans, it looks as though inflation didn't make just one universe and stop, but a huge and diverse population of parallel universes, perhaps realizing all possible options for what we used to think of as physical constants. Which would explain yet another mystery: the fact that many constants in our universe are so fine-tuned for life that if they changed by small amounts, life as we know it would be impossible—there would be no galaxies, or no atoms, say. Even though most of the parallel universes created by inflation are stillborn, there will be some where conditions are just right for life, and it's not surprising that this is where we find ourselves.

Inflation has given us an embarrassment of riches—and embarrassing it is. Because this infinity of universes has brought about the so-called measurement problem, which I view as the greatest crisis facing modern physics. Physics is all about predicting the future from the past, but inflation seems to sabotage this. Our physical world is clearly teeming with patterns and regularities, yet when we try quantifying them to predict the probability that something particular will happen, inflation always gives the same useless answer: infinity divided by infinity.

The problem is that whatever experiment you make, inflation predicts that there will be infinite copies of you obtaining each physically possible outcome in an infinite number of parallel universes, and despite years of teeth-grinding in the cosmology community, no consensus has emerged on how to extract sensible answers from these infinities. So, strictly speaking, we physicists are no longer able to predict anything at all. Our baby universe has grown into an unpredictable teenager.

This is so bad that I think a radical new idea is needed. Perhaps we need to somehow get rid of the infinite. Perhaps, like a rubber band, space can't be expanded ad infinitum without undergoing a big snap? Perhaps those infinite parallel universes get destroyed by some yet undiscovered process, or perchance they're, for some reason, mere mirages? The very deepest explanations provide not just answers but questions as well. I think inflation still has some explaining left to do.

KEPLER ET AL. AND THE NONEXISTENT PROBLEM

GINO SEGRÈ

Physicist, University of Pennsylvania; author, Ordinary
Geniuses: Max Delbruck, George Gamow, and the
Origins of Genomics and Big Bang Cosmology

In 1595, Johannes Kepler proposed a deep, elegant, and beautiful solution to the problem of determining the distance from the sun of the six then-known planets. Nesting (like Russian dolls) each of the five Platonic solids within a sphere, arranged in the proper order—octahedron, icosahedron, dodecahedron, tetrahedron, cube—he proposed that the succession of spherical radii would have the same relative ratios as the planetary distances. Of course the deep, elegant, and beautiful solution was also wrong, but then, as Joe E. Brown famously said at the conclusion of *Some Like It Hot,* "Nobody's perfect."

Two thousand years earlier, in a notion that would come to be described as the Harmony of the Spheres, Pythagoras had already sought a solution by relating those distances to the sites on a string where it needed to be plucked in order to produce notes pleasing to the ear. And almost 200 years after Kepler's suggestion, Johann Bode and Johann Titius offered, with no underlying explanation, a simple numerical formula that supposedly fit the distances in question. So we see that Kepler's explanation was neither the first nor the last attempt to explain the ratios of planetary-orbit radii, but in its linking of dynamics to geometry it remains, for me, the deepest, as well as being the simplest and most elegant.

In a strict sense, none of the three proposals is strictly wrong.

They are instead solutions to a problem that doesn't exist, for we now understand that the location of planets is purely accidental, a by-product of how the swirling disk of dust that circled our early sun evolved, under the force of gravity, into its present configuration. The realization that there was no problem came as our view expanded from one in which our planetary system was central to a far greater vision, in which it is one of an almost limitless number of such systems scattered throughout the vast numbers of galaxies comprising our universe.

I have been thinking about this because, together with many of my fellow theoretical physicists, I have spent a good part of my career searching for an explanation of the masses of the so-called elementary particles. But perhaps the reason it has eluded us is a proposal that is increasingly gaining credence—namely, that our visible universe is only a random example of an essentially infinite number of universes, all of which contain quarks and leptons with masses taking different values. It just happens that in at least one of those universes, the values allow for there being at least one star and one planet where creatures that worry about such problems live.

In other words, a problem we thought was central may once again have ceased to exist, as our conception of the universe has grown—in this case, been extended to one of many universes. If this is true, what grand vistas may lie before us in the future? I only hope that our descendants may have a much deeper understanding of these problems than we do and that they will smile at our feeble attempts to provide a deep, elegant, and beautiful solution to what they have recognized as a nonexistent problem.

HOW INCOMPATIBLE WORLDVIEWS CAN COEXIST

FREEMAN DYSON

Theoretical physicist, Institute for Advanced Study; author, A Many-Colored Glass: Reflections on the Place of Life in the Universe

The situation I am trying to explain is the existence side by side of two apparently incompatible pictures of the universe. One is the classical picture of our world as a collection of things and facts that we can see and feel, dominated by universal gravitation. The other is the quantum picture of atoms and radiation that behave in an unpredictable fashion, dominated by probabilities and uncertainties. Both pictures appear to be true, but the relationship between them is a mystery.

The orthodox view among physicists is that we must find a unified theory that includes both pictures as special cases. The unified theory must include a quantum theory of gravitation, so that particles called gravitons must exist, combining the properties of gravitation with quantum uncertainties.

I am looking for a different explanation of the mystery. I ask whether a graviton, if it exists, could conceivably be observed.

I do not know the answer to this question, but I have one piece of evidence that the answer may be no. The evidence is the behavior of one piece of apparatus—the gravitational-wave detector called LIGO (Laser Interferometer Gravitational-Wave Observatory) now operating in Louisiana and Washington State. The way LIGO works is to measure very accurately the distance between two mirrors by bouncing light from one to the other. When a gravitational wave comes by, the distance between the two mirrors will change very slightly. Because of ambient and instrumental noise, the actual LIGO detectors can only detect waves far stronger than a single graviton. But even in a totally quiet universe, I can answer the question of whether an ideal LIGO detector could detect a single graviton. The answer is no. In a quiet universe, the limit to the accuracy of measurement of distance is set by the quantum uncertainties in the positions of the mirrors. To make the quantum uncertainties small, the mirrors must be heavy. A simple calculation, based on the known laws of gravitation and quantum mechanics, leads to a striking result. To detect a single graviton with a LIGO apparatus, the mirrors must be exactly so heavy that they will attract each other with irresistible force and collapse into a black hole. In other words, Nature herself forbids us to observe a single graviton with this kind of apparatus.

I propose as a hypothesis, based on this single thought experiment, that single gravitons may be unobservable by any conceivable apparatus. If this hypothesis is true, it would imply that theories of quantum gravity are untestable and therefore scientifically meaningless. The classical universe and the quantum universe could then live together in peaceful coexistence. No incompatibility between the two pictures could ever be demonstrated. Both pictures of the universe could be true, and the hope of a unified theory could turn out to be an illusion.

IMPOSSIBLE INEXACTNESS

SATYAJIT DAS

Derivatives expert, risk-management consultant; author, Extreme
Money: The Masters of the Universe and the Cult of Risk

Inexactness is an end often seen as the beginning. Its profound
beauty transects science, mathematics, method, philosophy, linguistics, and faith.

In 1927, Werner Heisenberg showed that uncertainty is inherent in quantum mechanics. It is impossible to simultaneously measure certain properties of a particle—position and momentum. In
the quantum world, matter can take the form of either particles or
waves. Fundamental elements are neither particles nor waves but
can behave as either and are merely different theoretical ways of
picturing the quantum world.

Inexactness marks an end to certainty. As we seek to measure
one property more precisely, the ability to measure the other property is undermined. The act of measurement negates elements of
our knowledge of the system.

Inexactness undermines scientific determinism, implying that
human knowledge about the world is always incomplete, uncertain, and highly contingent.

Inexactness challenges causality. As Heisenberg observed:
"Causality law has it that if we know the present, then we can
predict the future. Be aware: In this formulation, it is not the consequence but the premise that is false. As a matter of principle, we
cannot know all determining elements of the present."

Inexactness questions methodology. Experiments can prove
only what they are designed to prove. Inexactness is a theory based
on the practical constraints of measurement.

Inexactness and quantum mechanics challenge faith as well as concepts of truth and order. They imply a probabilistic world of matter, where we cannot know anything with certainty but only as a possibility. It removes the Newtonian elements of space and time from any underlying reality. In the quantum world, mechanics are understood as a probability without any causal explanation.

Albert Einstein refused to accept that positions in spacetime could never be completely known and quantum probabilities did not reflect any underlying causes. He did not reject the theory but the lack of reason for an event. Writing to Max Born, he famously stated, "I, at any rate, am convinced that He [God] does not throw dice." But as Stephen Hawking later remarked, in terms that Heisenberg would have recognized, "Not only does God play dice, but . . . he sometimes throws them where they cannot be seen."

Allusive and subtle, the power of inexactness draws on its metaphorical property, which has allowed it to penetrate diverse fields, such as art theory, financial economics, and even popular culture. At one level, Heisenberg's uncertainty principle is taken to mean that the act of measuring something changes what is observed. But at another level, intentionally or unintentionally, Heisenberg is saying something about the nature of the entire system—the absence of absolute truths and the limits to our knowledge.

Inexactness is linked with various philosophical constructs. Søren Kierkegaard differentiated between objective truths and subjective truths. Objective truths are filtered and altered by our subjective truths, recalling the interaction between observer and event central to Heisenberg's theorem.

Inexactness is related to linguistic philosophies. In the *Tractatus Logico-Philosophicus*, Ludwig Wittgenstein anticipates inexactness, arguing that the structure of language provides the limits of thought and what can be said meaningfully.

The deep ambiguity of inexactness manifests itself in other ways: the controversy over Heisenberg's personal history. In 1941, during the Second World War, Heisenberg and Niels Bohr, his former teacher, met in occupied Denmark. In Michael Frayn's 1998 play *Copenhagen*, Margrethe, Bohr's wife, poses the essential question, which is debated in the play: "Why did he [Heisenberg] come to Copenhagen?" The play repeats their meeting three times, each with different outcomes. As Heisenberg, the character, states: "No one understands my trip to Copenhagen. Time and time again I've explained it. To Bohr himself, and Margrethe. To interrogators and intelligence officers, to journalists and historians. The more I've explained, the deeper the uncertainty has become."

In his 1930 text *The Principles of Quantum Mechanics*, Paul Dirac contrasted the Newtonian world and the quantum one: "It has become increasingly evident . . . that nature works on a different plan. Her fundamental laws do not govern the world as it appears in our mental picture in any direct way, but instead they control a substratum of which we cannot form a mental picture without introducing irrelevancies."

There was a world before Heisenberg and his uncertainty principle. There is a world after Heisenberg. They are the same world, but they are different.

THE NEXT LEVEL OF FUNDAMENTAL MATTER?

HAIM HARARI

Theoretical physicist; former president, Weizmann Institute of Science; author, A View from the Eye of the Storm

A scientific idea may be elegant. It may also be correct. If you must choose, choose correct. But it's always better to have both.

"Elegant" is in the eye of the beholder. "Correct" is decided by the ultimate judge of science, Mother Nature, speaking through the results of experiments. Unlike the standard TV talent contests, neither "elegant" nor "correct" can be determined by a vote of the public or by a panel of sneering judges. But the feeling that an idea is elegant often depends on the question being asked.

All matter consists of six types of quarks and six types of leptons, with seemingly random unexplained mass values, spanning more than ten orders of magnitude. No one knows why, within these twelve building blocks, the same pattern repeats itself three times. Some of these objects may also convert into each other, under certain circumstances, by unexplained rates called "mixing angles." The twenty-odd values of these rates and masses seem to have been arbitrarily chosen by someone (Nature or God). This is what the Standard Model of Particle Physics tells us. Is this elegant? It does not seem so.

But the fact that mountains and snakes, oceans and garbage, people and computers, hamburgers and stars, diamonds and elephants, and everything else in the universe are all made of only a dozen types of fundamental objects is truly mind boggling. That is exactly what that same Standard Model says. So is it elegant? Very much so.

My great hope is that nature is actually even more elegant. The twelve fundamental quarks and leptons and their antiparticles all have electric charges 0, ⅓, ⅔, and 1 or the negative values of the same numbers. Each value repeats exactly three times.

There is no satisfactory explanation for many questions: Why are all charges multiples of ⅓ of the electron charge? Why does each value between 0 and 1 appear on the list, and do so the same number of times? Why do they never acquire more than three doses of that quantity? Why does the same entire pattern repeat itself three times? Why do the leptons always have integer charges and the quarks non-integers? Why are quark charges and lepton charges at all related to each other by simple ratios?

The fact that mosquitoes, chairs, and tomato juice are all electrically neutral results from the unexplained equality of the magnitudes of the electric charges of protons and electrons, causing atoms to be neutral. This follows from the quark charges having precise simple ratios to the lepton charges. But why doesn't the electron have a charge of, say, 0.8342 of that of the proton? Why do they have exactly the same charge value?

An elegant explanation for these puzzles would appear if all quarks and leptons (and therefore all matter in the universe) consisted of only two building blocks, one with electric charge of ⅓ that of the electron and one without electric charge. Then all combinations of such three objects might exactly create the known pattern of quarks and leptons and neatly answer the above questions. The bizarre list of masses and conversion rates of the quarks and leptons would still remain unexplained but would be relegated to a level of discussion of understanding the dynamical forces binding the two more fundamental basic objects into a variety of compounds, rather than as a God-given or nature-given list of more than twenty free fundamental parameters.

An elegant explanation? Certainly. Correct? Not necessarily, as far as we know now. But you can never prove that particles are not made of more fundamental objects. This may well be discovered in the future without contradicting any current data, especially if the new structure is revealed only at smaller distances and higher energies than anything we have seen so far, or if it obeys a strange new set of basic physics rules. Needless to say, such a simple hypothesis needs to tackle many additional issues, some of which it does beautifully, while in others it fails badly. That may be the partly justified reason for the general negative attitude of most particle physicists to this simple explanation.

I find the idea of creating the entire universe from just two types of building blocks (which I call *Rishons*, or primaries) an elegant and enticing explanation of many observed facts. The book of Genesis starts with a universe that is "formless and void," or, in the original Hebrew, "*Tohu Vavohu*." What better notations for the two fundamental objects than T (*Tohu*, "formless") and V (*Vohu*, "void"), and then each quark or lepton would consist of a different combination of three such *Rishons*, like TTV or TTT. This may remain forever as an elegant but incorrect idea, or it may be revealed one day as the next level of the structure of matter, following the atom, the nucleus, the proton, and the quark. Ask Mother Nature. She understands both "elegant" and "correct," but she is not yet telling.

OBSERVERS OBSERVING

ROBERT PROVINE

Neuroscientist and psychologist, University of Maryland; author,
Curious Behavior: Yawning, Laughing, Hiccupping, and Beyond

The request for a favorite deep, elegant, and beautiful explanation left me a bit cold. "Deep," "elegant," and "beautiful" are aesthetic qualities I associate more with experience and process than explanation, especially that of the observer observing. Observation is the link between all empirical sciences and the reason physicists were among the founders of experimental psychology. The difference between psychology and physics is one of emphasis; both involve the process of observers observing. Physics stresses the observed, psychology the observer. As horrifying as this may be to hyperempiricists, who neglect the observer, physics is necessarily the study of the behavior of physicists, biology the study of biologists, and so on.

Decades ago, I discussed this issue with John Wheeler, who found it obvious, noting that a major limit on cosmology is the cosmologist. When students in my course on Sensation and Perception hear me say that we're engaging in the study of everything, I'm absolutely serious. In many ways, the study of sensation and perception is the most basic and universal of sciences.

My passion for observation is aesthetic as well as scientific. My most memorable observations are of the night sky. Others may name the discovery of a *T. rex* fossil or the sound of birdsong on a perfect spring day. To see better and deeper, I build telescopes, large and small. I like my photons fresh, not collected by CCD or analyzed by computer. I want to encounter the cosmos head on, letting it wash over my retina. My profession of neuroscience provides its own observational adventures, including the unique opportunity to close the circle by investigating the neurological mechanism through which the observer observes and comes to know the cosmos.

GENES, CLAUSTRUM, AND CONSCIOUSNESS

V. S. RAMACHANDRAN

Neuroscientist; professor & director, Center for Brain and Cognition, University of California–San Diego; author, The Tell-Tale Brain

What's my favorite elegant idea? The elucidation of DNA's structure is surely the most obvious, but it bears repeating. I'll argue that the same strategy used to crack the genetic code might prove successful in cracking the "neural code" of consciousness and self. It's a long shot, but worth considering.

The ability to grasp analogies, and to see the difference between deep and superficial ones, is a hallmark of many great scientists. Francis Crick and James Watson were no exception. Crick himself cautioned against the pursuit of elegance in biology, given that evolution proceeds happenstantially. "God is a hacker," he said, adding (according to my colleague Don Hoffman), "Many a young biologist has slit his own throat with Occam's razor." Yet his own solution to the riddle of heredity ranks with natural selection as biology's most elegant discovery. Will a solution of similar elegance emerge for the problem of consciousness?

It is well known that Crick and Watson unraveled the double-helical structure of the DNA molecule: two twisting complementary strands of nucleotides. Less well known is the chain of events culminating in this discovery.

First, Mendel's laws dictated that genes are particulate (a first approximation, still held to be accurate). Then Thomas Morgan showed that fruit flies zapped with X-rays became mutants with punctate changes in their chromosomes, yielding the clear conclusion that the chromosomes are where the action is. Chromosomes are composed of histones and DNA; as early as 1928, the British bacteriologist Fred Griffith showed that a harmless species of bacterium, upon incubation with a heat-killed virulent species, changes into the virulent species. This was almost as startling as a pig walking into a room with a sheep and two sheep emerging. Later, Oswald Avery showed that DNA was the transformative principle here. In biology, knowledge of structure often leads to knowledge of function—one need look no further than the whole of medical history. Inspired by Griffith and Avery, Crick and Watson realized that the answer to the problem of heredity lay in the structure of DNA. Localization was critical, as, indeed, it may prove to be for brain function.

Crick and Watson didn't just describe DNA's structure, they explained its significance. They saw the analogy between the complementarity of molecular strands and the complementarity of parent and offspring—why pigs beget pigs and not sheep. At that moment, modern biology was born. There are similar correlations between brain structure and mind function, between neurons and consciousness. (I'm stating the obvious here only because there are some philosophers, called "new mysterians," who believe the opposite.)

After his triumph with heredity, Crick turned to what he called the "second great riddle" in biology—consciousness. There were many skeptics. I remember a seminar Crick gave on consciousness at the Salk Institute here in La Jolla. He'd barely started when a gentleman in attendance raised a hand and said, "But Dr. Crick, you haven't even bothered to *define* the word 'consciousness' before embarking on this." Crick's response was memorable: "I'd remind you that there was never a time in the history of biology when a bunch of us sat around the table and said, 'Let's first *define* what we mean by life.' We just went out there and discovered what it was—a double helix. We leave matters of semantic hygiene to you philosophers."

Crick did not, in my opinion, succeed in solving consciousness (whatever it might mean). Nonetheless, he was headed in the right direction. He had been richly rewarded earlier in his career for grasping the analogy between biological complementarities, the notion that the structural logic of the molecule dictates the functional logic of heredity. Given his phenomenal success using the strategy of structure-function analogy, it is hardly surprising that he imported the same style of thinking to study consciousness. He and his colleague Christof Koch did so by focusing on a relatively obscure structure called the claustrum.

The claustrum is a thin sheet of cells underlying the insular cortex of the brain, one on each hemisphere. It is histologically more homogeneous than most brain structures, and unlike most brain structures (which send and receive signals to and from a small subset of other structures), the claustrum is reciprocally connected with almost every cortical region. The structural and functional streamlining might ensure that when waves of information come through the claustrum, its neurons will be exquisitely sensitive to the timing of the inputs.

What does this have to do with consciousness? Instead of focusing on pedantic philosophical issues, Crick and Koch began with their naive intuitions. "Consciousness" has many attributes—continuity in time; a sense of agency or free will; recursiveness, or "self-awareness," etc. But one attribute that stands out is subjective unity: You experience all your diverse sense impressions, thoughts, willed actions, and memories as a unity—not as jittery or fragmented. This attribute of consciousness, with the accompanying sense of the immediate present, or the "here and now," is so obvious that we don't usually think about it; we regard it as axiomatic.

So a central feature of consciousness is its unity—and here is a brain structure that sends and receives signals to and from practically all other brain structures, including the right parietal (involved in polysensory convergence and embodiment) and the anterior cingulate (involved in the experience of "free will"). Thus the claustrum seems to unify everything anatomically, and consciousness does so mentally. Crick and Koch recognized that this might not be a coincidence: The claustrum may be central to consciousness—indeed, it may embody the idea of the Cartesian theater, taboo among philosophers—or at least be the conductor of the orchestra. It is this kind of childlike reasoning that often leads to great discoveries. Obviously, such analogies don't replace rigorous science, but they're a good place to start. Crick and Koch's idea may be right or wrong, but it's elegant. If it's right, they have paved the way to solving one of the great mysteries of biology. Even if it's wrong, students entering the field would do well to emulate their style. Crick was right too often to ignore.

I visited him at his home in La Jolla in July of 2004. He saw me to the door as I was leaving and, as we parted, gave me a sly, conspiratorial wink: "I think it's the claustrum, Rama. That's where the secret is." A week later, he passed away.

OVERLAPPING SOLUTIONS

DAVID M. EAGLEMAN
Neuroscientist, Baylor College of Medicine; author,
Incognito: The Secret Lives of the Brain

The elegance of the brain lies in its inelegance. For centuries, neuroscience attempted to neatly assign labels to the various parts of the brain: This is the area for language, this for morality, this for tool use, color detection, face recognition, and so on. The search for an orderly brain map started off as a viable endeavor but turned out to be misguided.

The deep and beautiful trick of the brain is more interesting: It possesses multiple, overlapping ways of dealing with the world. It is a machine built of conflicting parts. It is a representative democracy that functions by *competition* among parties who all believe they know the right way to solve the problem.

As a result, we can get mad at ourselves, argue with ourselves, curse at ourselves, and contract with ourselves. We can feel conflicted. These sorts of neural battles lie behind marital infidelity, relapses into addiction, cheating on diets, breaking of New Year's resolutions—all situations in which some parts of a person want one thing and other parts another.

These are things that modern machines simply do not do. Your car cannot be conflicted about which way to turn: It has one steering wheel commanded by one driver, and it follows directions without complaint. Brains, on the other hand, can be of two minds, and often many more. We don't know whether to turn toward the cake or away from it, because there are several sets of hands on the steering wheel of behavior.

Take memory. Under normal circumstances, memories of daily

events are consolidated by an area of the brain called the hippo-campus. But in frightening situations—such as a car accident or a robbery—another area, the amygdala, also lays down memories along an independent, secondary memory track. Amygdala memories have a different quality to them: They are difficult to erase and they can return in "flash-bulb" fashion—a common description of rape victims and war veterans. In other words, there is more than one way to lay down memory. We're talking not about memories of different events but about different memories of the same event. According to the unfolding picture, there may be even more than two factions involved, all writing down information and later competing to tell the story. The unity of memory is an illusion.

Consider the different systems involved in decision making: Some are fast, automatic, and below the surface of conscious awareness; others are slow, cognitive, and conscious. And there's no reason to assume that there are only two systems; there may well be a spectrum. Some networks in the brain are implicated in long-term decisions, others in short-term impulses—and there may be a fleet of medium-term biases as well.

Attention, also, has recently come to be understood as the end result of multiple, competing networks, some for focused, dedicated attention to a specific task and others for monitoring broadly (vigilance). They are always locked in competition to steer the actions of the organism. Even basic sensory functions like the detection of motion appear now to have been reinvented multiple times by evolution. This provides the perfect substrate for a neural democracy.

On a larger anatomical scale, the two hemispheres of the brain, left and right, can be understood as overlapping systems that compete. We know this from patients whose hemispheres are disconnected: They essentially function with two independent brains. For example, put a pencil in each hand and they can simultaneously draw incompatible figures, such as a circle and a triangle. The two hemispheres function differently in the domains of language, abstract thinking, story construction, inference, memory, gambling strategies, and so on. They constitute a team of rivals: agents with the same goals but slightly different ways of going about them.

To my mind, this elegant solution to the mysteries of the brain should change the goal for aspiring neuroscientists. Instead of spending years advocating for your favorite solution, the mission should evolve into elucidating the various overlapping solutions: how they compete, how the union is held together, and what happens when things fall apart.

Part of the importance of discovering elegant solutions is capitalizing on them. The neural-democracy model may be just the thing to dislodge artificial intelligence. We human programmers still approach a problem by assuming there's a best way to solve it or there's a way it *should* be solved. But evolution does not solve a problem and then check it off the list. Instead, it ceaselessly reinvents programs, each with overlapping and competing approaches. The lesson is to abandon the question "What's the cleverest way to solve this problem?" in favor of "Are there multiple, overlapping ways to solve this problem?" That will be the starting point in ushering in a fruitful new age of elegantly inelegant computational devices.

OUR BOUNDED RATIONALITY

MAHZARIN BANAJI
*Richard Clarke Cabot Professor of Social Ethics,
Department of Psychology, Harvard University*

Explanations that are extraordinary both analytically and aesthetically share, among others, these properties: (a) They are often simpler compared with what was received wisdom, (b) they point to the truer cause as being something quite removed from the phenomenon, and (c) they make you wish you'd come upon the explanation yourself.

Those of us who attempt to understand the mind have a unique limitation to confront: The mind is the thing doing the explaining; the mind is also the thing to be explained. Distance from one's own mind, distance from attachments to the specialness of one's species or tribe, getting away from introspection and intuition (not as hypothesis generators but as answers and explanations) are all especially hard to achieve when what we seek to do is explain our own minds and those of others of our kind.

For this reason, my candidate for the most deeply satisfying explanation of recent decades is the idea of bounded rationality. The idea that human beings are smart compared to other species but not smart enough by their own standards, including behaving in line with basic axioms of rationality, is now a well-honed observation with a deep empirical foundation.

The cognitive scientist and Nobel laureate in economics Herbert Simon put one stake in the ground through the study of information processing and artificial intelligence, showing that people and organizations alike adopt principles of behaviors such as "satisficing" that constrain them to decent but not the best decisions.

The second stake was placed by Daniel Kahneman and Amos Tversky, who showed the stunning ways in which even experts are error-prone, with consequences not only for their own welfare but also for that of their societies.

The view of human nature that has evolved over the past four decades has systematically changed the explanation for who we are and why we do what we do. We are error-prone in the unique ways in which we are, the explanation goes, not because we have malign intent but because of the evolutionary basis of our mental architecture—the way in which we learn and remember information, the way in which we are affected by those around us, and so on. The reason we are boundedly rational is because the information space in which we must do our work is large compared to the capacities we have, including severe limits on our conscious awareness and our ability to control our behavior and act in line with our own intentions.

We can also look at the compromise of ethical standards: Again, the story is the same; that is, it's not the intention to harm that's the problem. Rather, the explanation lies in such sources as the manner in which some information plays a disproportionate role in our decision making, the ability to generalize or overgeneralize, and the commonness of wrongdoing that typifies daily life. These are the more potent causes of the ethical failures of individuals and institutions.

The idea that bad outcomes result from limited minds that cannot store, compute, or adapt to the demands of their environment is a radically different explanation of our capacities and therefore our nature. Its elegance and beauty come from its emphasis on the ordinary and the invisible rather than on specialness and malign motives. This is not so dissimilar from another shift in explanation—from God to natural selection—and is likely to be equally resisted.

SWARM INTELLIGENCE

ROBERT SAPOLSKY

Professor of neurology and neurological sciences, Stanford
University; research associate, National Museums of Kenya; author,
Monkeyluv: And Other Essays on Our Lives as Animals

The obvious answer should be the double helix. With the incomparably laconic "It has not escaped our notice . . . ," it explained the very mechanism of inheritance. But the double helix doesn't do it for me. By the time I got around to high school biology, the double helix was ancient history, like pepper moths evolving or mitochondria as the powerhouses of the cell. Watson and Crick—as comforting, but as taken for granted, as Baskin and Robbins.

Then there's the work of Hubel and Wiesel, which showed that the cortex processes sensations with a hierarchy of feature extraction. In the visual cortex, for example, neurons in the initial layer each receive inputs from a single photoreceptor in the retina. Thus, when one photoreceptor is stimulated, so is "its" neuron in the primary visual cortex. Stimulate the adjacent photoreceptor, and the adjacent neuron activates. Basically, each of these neurons "knows" one thing—namely, how to recognize a particular dot of light. Groups of I-know-a-dot neurons then project onto single neurons in the second cortical layer. Stimulate a particular array of adjacent neurons in that first cortical layer, and a single second-layer neuron activates. Thus, a second-layer neuron knows one thing, which is how to recognize, say, a 45-degree-angle line of light. Then groups of I-know-a-line neurons send projections on to the next layer.

Beautiful, explains everything—just keep going, cortical layer upon layer of feature extraction, dot to line to curve to collection of curves, until the top layer, where a neuron would know one complex, specialized thing only, like how to recognize your grandmother. And it would be the same in the auditory cortex: first-layer neurons knowing particular single notes, second layer knowing pairs of notes, some neuron at the top that would recognize the sound of your grandmother singing along with Lawrence Welk.

It turns out, though, that things didn't quite work that way. There are few "grandmother neurons" in the cortex (although a 2005 *Nature* paper reported someone with a Jennifer Aniston neuron). The cortex can't rely too much on grandmother neurons, because that requires a gazillion more neurons to accommodate such inefficiency and overspecialization. Moreover, a world of nothing but grandmother neurons on top precludes making multimodal associations (for example, when seeing a particular Monet reminds you of croissants and Debussy's music and the disastrous date you had at an Impressionism show at the Met). Instead, we've entered the world of neural networks.

Which brings me to my selection, which is emergence and complexity, as represented by "swarm intelligence." Observe a single ant and it doesn't make much sense—walking in one direction, suddenly careening in another for no obvious reason, doubling back on itself. Thoroughly unpredictable. The same happens with two ants, with a handful of ants. But a colony of ants makes fantastic sense. Specialized jobs, efficient means of exploiting new food sources, complex underground nests with temperature regulated within a few degrees. And critically, there's no blueprint or central source of command—each individual ant has algorithms for its behaviors. But this is not the wisdom of the crowd, where a bunch of reasonably informed individuals outperform a single expert. The ants aren't reasonably informed about the big picture. Instead, the behavior algorithms of each ant consist of a few simple rules for interacting with the local environment and local ants. And out of this emerges a highly efficient colony.

Ant colonies excel at generating trails that connect locations in the shortest possible way, accomplished with simple rules about when to lay down a pheromone trail and what to do when encountering someone else's trail—approximations of optimal solutions to the Traveling Salesman problem. In "ant-based routing," simulations using virtual ants with similar rules can generate optimal ways of connecting the nodes in a network, something of great interest to telecommunications companies. It also applies to the developing brain, which must wire up vast numbers of neurons with vaster numbers of connections without constructing millions of miles of connecting axons. And migrating fetal neurons generate an efficient solution with a different version of ant-based routing.

A wonderful example is how local rules about attraction and repulsion (that is, positive and negative charges) allow simple molecules in an organic soup to occasionally form more complex ones. Life may have originated this way, without the requirement of bolts of lightning to catalyze the formation of complex molecules.

And why is self-organization so beautiful to my atheistic self? Because if complex adaptive systems don't require a blueprint, they don't require a Blueprint Maker. If they don't require lightning bolts, they don't require Someone hurtling lightning bolts.

LANGUAGE AND NATURAL SELECTION

KEITH DEVLIN
Executive director, H-STAR Institute, Stanford University; author,
The Man of Numbers: Fibonacci's Arithmetic Revolution

Not only does evolution by natural selection explain how we all got here and how we are and behave as we do, it can even explain (at least to my fairly critical satisfaction) why many people refuse to accept it and why even more people believe in an all-powerful Deity. But since other *Edge* respondents are likely to have natural selection as their favorite deep, elegant, and beautiful explanation (it has all three attributes, in addition to wide-ranging explanatory power), I'll home in on one particular instance: the explanation of how humans acquired language, by which I mean grammatical structure.

There is evidence to suggest that our ancestors developed effective means to communicate using verbal utterances starting at least 3 million years ago. But grammar is much more recent, perhaps as recent as 75,000 years ago. How did grammar arise?

Anyone who has traveled abroad knows that to communicate basic needs, desires, and intentions to people in your vicinity concerning objects within sight, a few referring words together with gestures suffice. The only grammar required is to occasionally juxtapose two words ("Me Tarzan, you Jane" being the information- [and innuendo-] rich classic example from Hollywood). Anthropologists refer to such a simple word-pairing communication system as protolanguage.

But to communicate about things not in the here-and-now, you need more. Effectively planning future joint activities needs pretty well all of grammatical structure, particularly if the planning involves more than two people—with even more demands made on the grammar if the plan requires coordination among groups not all present at the same place or time.

Given the degree to which human survival depends on our ability to plan and coordinate our actions and to collectively debrief after things go wrong so we avoid repeating our mistakes, it's clear that grammatical structure is hugely important to *Homo sapiens*. Indeed, many argue that it's our defining characteristic. But communication, while arguably the killer app for grammar, clearly cannot be what put it into the gene pool in the first place, and for a very simple reason. Since grammar is required in order for verbal utterances to convey ideas more complex than is possible with protolanguage, it comes into play only when the brain can form such ideas. These considerations lead to what is accepted (although not without opposition) as the Standard Explanation of language acquisition. In highly simplified terms, the Standard Explanation runs like this.

1. Brains (or the organs that became brains) first evolved to associate motor responses to sensory input stimuli.

2. In some creatures, brains became more complex, performing a mediating role between input stimuli and motor responses.

3. In some of those creatures, the brain became able to override automatic stimulus-response sequences.

4. In *Homo sapiens*, and to a lesser extent in other species, the brain acquired the ability to function off-line, effectively

running simulations of actions without the need for sensory input stimuli and without generating output responses.

Stage 4 is when the brain acquires grammar. What we call grammatical structure is in fact a descriptive/communicative manifestation of a mental structure for modeling the world.

As a mathematician, what I like about this explanation is that it also tells us where the brain got its capacity for mathematical thinking. Mathematical thinking is essentially another manifestation of the brain's simulation capacity, but in quantitative/relational/logical terms rather than descriptive/communicative.

As is usually the case with natural-selection arguments, it takes considerable work to flesh out the details of these simplistic explanations (and some days I'm less convinced than others about some aspects), but overall they strike me as about right. In particular, the mathematical story explains why doing mathematics carries with it an overpowering Platonic sense of reasoning not about abstractions but about real objects—at least "real" in a Platonic realm. At which point, the lifelong mathematics educator in me says I should leave the proof of that corollary as an exercise for the reader—so I shall.

COMMITMENT

RICHARD H. THALER

*Theorist, behavioral economics; director, Center for Decision
Research, Graduate School of Business, University of Chicago;
coauthor (with Cass R. Sunstein),* Nudge: Improving
Decisions About Health, Wealth, and Happiness

It is a fundamental principle of economics that a person is always better off if they have more alternatives to choose from. But this principle is wrong. There are cases when I can make myself better off by restricting my future choices and committing myself to a specific course of action.

The idea of commitment as a strategy is an ancient one. Odysseus famously had his crew tie him to the mast so he could listen to the sirens' songs without steering the ship into the rocks. Another classic is Cortez's decision to burn his ships upon arriving in South America, thus removing retreat as one of the options his crew could consider. But although the idea is an old one, we did not understand its nuances until Nobel laureate Thomas Schelling wrote his 1956 masterpiece: "An Essay on Bargaining."

It is well known that games such as the Prisoner's Dilemma work out if both players can credibly commit themselves to cooperating, but how can I convince you that I will cooperate when it is a dominant strategy for me to defect? (And if you and I are game theorists, you know that I know that you know that I know that defecting is a dominant strategy.) Schelling gives many examples of how this can be done, but here's my favorite: A Denver rehabilitation clinic whose clientele consisted of wealthy cocaine addicts offered a "self-blackmail" strategy. Patients were offered an opportunity to write a self-incriminating letter, which would be delivered if and only if the patient, who would be tested on a random schedule, was found to have used cocaine. Most patients would now have a very strong incentive to stay off drugs; they were committed.

Many of society's thorniest problems, from climate change to Middle East conflict, could be solved if the relevant parties could only find a way to commit themselves to some future course of action. They would be well advised to study Tom Schelling in order to figure out how to make that commitment.

TIT FOR TAT

JENNIFER JACQUET
Clinical assistant professor of environmental studies, NYU

Selfishness can sometimes seem like the best strategy. It is the rational response to the Prisoner's Dilemma, for instance, where each individual in a pair can either cooperate or defect, leading to four potential outcomes. No matter what the other person does, selfish behavior (defecting) always yields greater return. But if both players defect, both do worse than if they had cooperated. Yet when political scientist Robert Axelrod and his colleagues ran hundreds of rounds of the Prisoner's Dilemma expressed using a mathematical equation on a computer, the repetition of the game led to a different result.

Experts from a wide range of disciplines submitted 76 different game strategies for Axelrod to try out against each other—some of them very elaborate. Each strategy would play against all the others for 200 rounds. In the end, the strategy that received the highest score was also the simplest. *Tit for Tat*, an if-then strategy where the player cooperates on the first move and thereafter does what its partner does, was the winner. The importance of reciprocity to the evolution of cooperation was detected by humans but simulated and verified with machines.

This elegant explanation was then documented in living egoists with an elegant experiment. Evolutionary biologist Manfred Milinski noticed Tit for Tat behavior in his subjects, three-spined sticklebacks. When he watched a pair of these fish approach a predator, he observed four options: they could swim side by side, one could take the lead while the other followed closely behind (or vice versa), or they could both retreat. These four scenarios satisfied the four inequalities that define the Prisoner's Dilemma.

For the experiment, Milinski wanted to use pairs of stickle-backs, but they're impossible to train. So he placed in the tank a single stickleback and a set of mirrors that would act like two different types of companions. In the first treatment, a parallel mirror was used to simulate a cooperative companion that swam alongside the subject stickleback. In the second treatment, an oblique mirror system set at a 32-degree angle simulated a defect-ing partner—that is, as the stickleback approached the predator, the companion appeared to fall increasingly and uncooperatively behind. Depending on the mirror, the stickleback felt he was shar-ing the risk equally or increasingly going it alone.

When the sticklebacks were partnered with a defector, they preferred the safer half of the tank, farthest away from the preda-tor. But in the trials with the cooperating mirror, the sticklebacks were twice as likely to venture into the half of the tank closest to the predator. The sticklebacks were more adventurous if they had a sidekick. In nature, cooperative behavior translates to more food and more space, and therefore greater individual reproductive success. Contrary to predictions that selfish behavior or retreat was optimal, Milinski's observation that sticklebacks most often approached the predator together was in line with Axelrod's con-clusion that Tit for Tat was the optimal evolutionary strategy.

Milinski's evidence, published in 1987 in the journal *Nature*,* was the first to demonstrate that cooperation based on reciprocity definitely evolved among egoists, albeit small ones. A large body of research now shows that many biological systems, especially human societies, are organized around various cooperative strategies; the scientific meth-ods continue to become more and more sophisticated, but the original experiments and Tit For Tat strategy are beautifully simple.

---

* "TIT FOR TAT in sticklebacks and the evolution of cooperation," *Nature* 325, 433-5 (1987).

TRUE OR FALSE: BEAUTY IS TRUTH

JUDITH RICH HARRIS
Independent investigator and theoretician; author, The Nurture
Assumption: Why Children Turn Out the Way They Do

"Beauty is truth, truth beauty," said John Keats. But what did he know? Keats was a poet, not a scientist. In the world that scientists inhabit, truth is not always beautiful or elegant, though it may be deep. In fact, it's my impression that the deeper an explanation goes, the less likely it is to be beautiful or elegant.

In 1938, the psychologist B. F. Skinner proposed an elegant explanation of "the behavior of organisms" (the title of his first book), based on the idea that rewarding a response—he called it *reinforcement*—increases the probability that the same response will occur in the future. The theory failed, not because it was false (reinforcement generally does increase the probability of a response) but because it was too simple. It ignored innate components of behavior. It couldn't even handle all learned behavior. Much behavior is acquired or shaped through experience, but not necessarily by means of reinforcement. Organisms learn various things in various ways.

The theory of the modular mind is another way of explaining behavior—in particular, human behavior. The idea is that the human mind is made up of a number of specialized components called modules that work more or less independently. These modules collect various kinds of information from the environment and process it in different ways. They issue different commands—occasionally, conflicting commands. It's not an elegant theory; on the contrary, it's the sort of thing that would make Occam whip out his razor. But we shouldn't judge theories by asking them to compete in a beauty pageant. We should ask whether they can explain more, or explain better, than previous theories could. The modular theory can explain, for example, the curious effects of brain injuries. Some abilities may be lost while others are spared, with the pattern differing from one patient to another.

More to the point, the modular theory can explain some of the puzzles of everyday life. Consider intergroup conflict. The Montagues and the Capulets hated each other; yet Romeo (a Montague) fell in love with Juliet (a Capulet). How can you love a member of a group yet go on hating that group? The answer is that two separate mental modules are involved. One deals with groupness (identification with one's group and hostility toward other groups), the other specializes in personal relationships. Both modules collect information about people, but they do different things with the data. The groupness module draws category lines and computes averages within categories; the result is called a stereotype. The relationship module collects and stores detailed information about specific individuals. It takes pleasure in collecting this information, which is why we love to gossip, read novels and biographies, and watch political candidates unravel on our TV screens. No one has to give us food or money to get us to do these things, or even administer a pat on the back, because collecting the data is its own reward.

The theory of the modular mind is not beautiful or elegant. But not being a poet, I prize truth above beauty.

ERATOSTHENES AND THE MODULAR MIND

DAN SPERBER

Social and cognitive scientist; director, International Cognition and Culture Institute; coauthor (with Deirdre Wilson), Meaning and Relevance

Eratosthenes (276–195 B.C.), the head of the famous Library of Alexandria in Ptolemaic Egypt, made groundbreaking contributions to mathematics, astronomy, geography, and history. He also argued against dividing humankind into Greeks and "Barbarians." What he is remembered for, however, is having provided the first correct measurement of the circumference of the Earth (a story well told in Nicholas Nicastro's recent book, *Circumference*). How did he do it?

Eratosthenes had heard that every year, on a single day at noon, the sun shone directly to the bottom of an open well in the town of Syene (now Aswan). This meant that the sun was then at the zenith. For that, Syene had to be on the Tropic of Cancer and the day had to be the summer solstice (our June 21). He knew how long it took caravans to travel from Alexandria to Syene and, on that basis, estimated the distance between the two cities to be 5,014 stades. He assumed that Syene was due south on the same meridian as Alexandria. Actually, in this he was slightly mistaken—Syene is somewhat to the east of Alexandria—and also in assuming that Syene was right on the Tropic. But, serendipitously, the effect of these two mistakes canceled each other out. He understood that the sun was far enough away to treat as parallel the rays of the sun that reach the Earth. When the sun was at the zenith in Syene, it had to be south of the zenith in the more northern Alexandria. By how much? He measured the length of the shadow cast by an obelisk located in front of the Library (says the story—or cast by some other, more convenient vertical object), and—even without trigonometry that had yet to be developed—he could determine that the sun was at an angle of 7.2° south of the zenith. That very angle, he understood, measured the curvature of the Earth between Alexandria and Syene (see the figure). Since 7.2° is a fiftieth of 360°, Eratosthenes could then, by multiplying the distance between Alexandria and Syene by 50, calculate the circumference of the Earth. The result, 252,000 stades, is 1 percent shy of the modern measurement of 40,008 km.

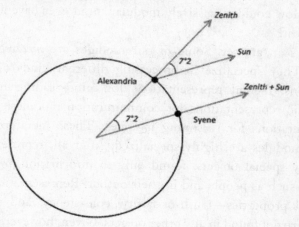

Eratosthenes brought together apparently unrelated pieces of evidence (the pace of caravans, the sun shining to the bottom of a well, the length of the shadow of an obelisk), assumptions (the sphericity of the Earth, its distance from the sun), and mathematical tools to measure a circumference he could neither see nor survey but only imagine. His result is simple and compelling. The way he reached it epitomizes human intelligence at its best.

Jerry Fodor (whose contribution to modern philosophy of mind is second to none) might well use this intellectual prowess as a perfect illustration of the way the central systems of our mind operate. They are, he claims, "isotropic," in the sense that any belief or evidence is relevant to the evaluation of any new hypothesis, and "Quinean" (after the philosopher Willard Van Orman Quine), in the sense that all our beliefs are part of a single integrated system. This contrasts with the view (which I have contributed to developing) that the mind is wholly made up of specialized "modules," each attending to a specific cognitive domain or task, and that our mental activity results from the complex interactions (complementarities, competitions . . .) among these modules. Doesn't, however, the story of Eratosthenes prove that Fodor's view is right? How could a massively modular mind ever have achieved such a feat?

Here's an answer. Some of our modules are *metarepresentational*. They specialize in processing different kinds of representations: mental representations for mind-reading modules; linguistic representations for communication modules; abstract representations for reasoning modules. These metarepresentational modules are highly specialized. After all, representations are very special objects, found only in information-processing devices such as people and in their output. Representations have original properties—truth-or-falsity, consistency, and so on—which are not found in any other objects. Given, however, that the representations these metarepresentational modules process may themselves be about just anything, they provide a kind of *virtual domain-generality*. Hence the illusion that metarepresentational thinking is truly general and nonspecialized.

Eratosthenes, I am suggesting, was not thinking concretely about the circumference of the Earth (in the way he might have been thinking concretely about the distance from the Library to the Palace in Alexandria). Rather, he was thinking about a challenge posed by the quite different estimates of the circumference of the Earth that had been offered by other scholars at the time. He was thinking about various mathematical principles and tools that could be brought to bear on the issue. He was thinking of the evidential use that could be made of sundry observations and reports. He was aiming at finding a clear and compelling solution, a convincing argument. In other terms, he was thinking about objects *of a single kind*—representations—and looking for a new way to put them together. In doing so, he was inspired by others and aiming at others. His intellectual feat makes sense only as a particularly remarkable link in a social-cultural chain of mental and communicational events. To me, it is a stunning illustration not of the solitary functioning of the individual mind but of the powers of socially and culturally extended modular minds.

DAN SPERBER'S EXPLANATION OF CULTURE

CLAY SHIRKY

Social media researcher; arts professor, NYU Tisch School of the Arts' Interactive Telecommunications Program (ITP); author, Cognitive Surplus: How Technology Makes Consumers into Collaborators

Why do members of a group of people behave the same way? Why do they behave differently from other groups living nearby? Why are those behaviors so stable over time? Alas, the obvious answer— cultures are adaptations to their environments—doesn't hold up. Multiple adjacent cultures along the Indus, the Euphrates, the Upper Rhine have differed in language, dress, and custom, despite existing side-by-side in almost identical environments.

Something happens to keep one group of people behaving in a certain set of ways. In the early 1970s, both E. O. Wilson and Richard Dawkins noticed that the flow of ideas in a culture exhibited similar patterns to the flow of genes in a species—high flow within the group but sharply reduced flow between groups. Dawkins' response was to assume a hypothetical unit of culture called the meme, though he also made its problems clear: With genetic material, perfect replication is the norm and mutations rare. With culture, it is the opposite; events are misremembered and then misdescribed, quotes are mangled, even jokes (pure meme) vary from telling to telling. The gene/meme comparison remained, for a generation, an evocative idea of not much analytic utility.

Dan Sperber has, to my eye, cracked this problem. In a slim 1996 volume modestly titled *Explaining Culture*, he outlined a theory of culture as the residue of the epidemic spread of ideas. In this model, there is no meme, no unit of culture separate from the blooming, buzzing confusion of transactions. Instead, all cultural transmission can be reduced to one of two types: making a mental representation public, or internalizing a mental version of a public presentation. As Sperber puts it, "Culture is the precipitate of cognition and communication in a human population."

Sperber's two primitives—externalization of ideas, internalization of expressions—give us a way to think of culture not as a big container that people inhabit but as a network whose traces, drawn carefully, let us ask how the behaviors of individuals create larger, longer-lived patterns. Some public representations are consistently learned and then reexpressed and relearned—Mother Goose rhymes, tartan patterns, and peer review have all survived for centuries. Others move from ubiquitous to marginal in a matter of years—pet rocks, "The Piña Colada Song." Still others thrive only within a subcultural domain—cosplay, Civil War reenactments. (Indeed, a subculture is simply a network of people who traffic in particular representations—representations that are mostly inert in the larger culture.)

With Sperber's network-tracing model, culture is best analyzed as an overlapping set of transactions rather than as a container or a thing or a force. Given this, we can ask detailed questions about which private ideas are made public where, and we can ask when and how often those public ideas take hold in individual minds.

Rather than arguing about whether the sonnet is still a vital part of Western culture, for example, Sperber makes it possible to ask instead, "Which people have mental representations of individual sonnets, or of the sonnet as an overall form? How often do they express those representations? How often do others remember those expressions?" Understanding sonnet-knowing becomes a network-analysis project, driven by empirical questions about how widespread, detailed, and coherent the mental representations of sonnets are. Cultural commitment to sonnets and Angry Birds and American exceptionalism and the theory of relativity can all be placed under the same lens.

This is what is so powerful about Sperber's idea: Culture is a giant, asynchronous network of replication, ideas turning into expressions, which turn into other, related ideas. Sperber also allows us to understand why persistence of public expression can be so powerful. When I sing "Camptown Races" to my son, he internalizes his own (slightly different) version. As he learns to read sheet music, however, he gains access to a much larger universe of such representations; Beethoven is not around to hum "Für Elise" to him, but through a set of agreed-on symbols (themselves internalized as mental representations) Beethoven's public representations can be internalized centuries later.

Sperber's idea also suggests that increased access to public presentation of ideas will increase the dynamic range of culture overall. Some publicly available representations will take hold among the widest possible group of participants in history, in both absolute numbers and as a percentage of the human race. (Consider, for example, the number of people who can now understand the phrase "That's killing two pigs with one bird.") It is this globally wired possibility for global cultural imitation that evolutionary theorist Mark Pagel worries about when he talks about the Internet enabling "infinite stupidity."

At the same time, it has never been easier for members of possible subcultures to find one another and create their own public representations at much lower cost, longer life, and greater reach than ordinary citizens have ever been able to. The January 25, 2011, protests in Egypt hijacked the official public representation of that day as National Police Day; this was possible only because the dissidents could create alternate public representations on a scale similar to the Egyptian state.

Actual reductionism—the interpretation of a large number of effects using a small number of causes—is rare in the social sciences, but Sperber has provided a framework for dissolving large and vague questions about culture into a series of tractable research programs. Most of the empirical study of the "precipitate of cognition and communication" is still in the future, but I can't think of another current idea in the social sciences offering that degree of explanatory heft.

METAREPRESENTATIONS EXPLAIN HUMAN UNIQUENESS

HUGO MERCIER

Psychologist; cognitive scientist; postdoc, University of Neuchâtel

Humans alone fluently understand the mental states of others. Humans alone rely on an open-ended system of communication. Humans alone ponder the reasons for their beliefs. For each of these feats, and for others too, humans rely on their most special gift: the ability to represent representations—that is, to form metarepresentations. Hidden behind such mundane thoughts as "Mary believes that Paul believes that it's going to rain" is the explanation of human uniqueness.

There are two ways to represent representations, one immensely powerful, the other rather clumsy. The clumsy way is to create a new representation for every representation that needs to be represented. Using such a device, Mary would have to form a representation "Paul believes that it's going to rain" completely independent of her representation "it's going to rain." She would then have to learn anew all of the inferences that can be drawn from "Paul believes that it's going to rain," such as the negative impact on Paul's willingness to go for a jog or the increased probability that he will fetch an umbrella. This cumbersome process would have to be repeated *for each new representation* that Mary wishes to attribute, from "Peter thinks the weather looks lovely" to "Ruth is afraid that the Dow Jones is going to crash tomorrow." Such a process could not possibly account for humans' amazing abilities to attribute just about any thought to other people. How then can we account for these skills?

The explanation is that we use our own representations to attribute thoughts to others. When Mary wants to attribute to Paul the belief "it's going to rain," she simply uses her representation "it's going to rain" and embeds it in a metarepresentation: "Paul thinks that 'it's going to rain.'" Because the same representation is used, Mary can take advantage of the inferences she could draw from "it's going to rain" to draw inferences from "Paul believes that 'it's going to rain.'" This trick opened for humans the doors to an unparalleled understanding of their social environment.

Most of the beliefs we form about others are derived from communication: People keep telling us what they believe, want, desire, fear, love. Here again, metarepresentations play a crucial role, since understanding language requires going from utterances— "It's going to rain"—to metarepresentations—"Paul means that 'it will soon rain here.'"

Mentalizing—attributing thoughts to others—and communicating are the best-known uses of metarepresentations, but they are not the only ones. Metarepresentations are also essential for us to be able to think about reasons. Specific metarepresentations are relied on when people produce and evaluate arguments, as in: "Mary thinks 'it's going to rain' is a good argument for 'We shouldn't go out.'" Again, Mary uses her representation "it's going to rain," but instead of attributing it to someone else, she represents its strength as a reason to accept a given conclusion.

Several other properties of representations can be represented, from their aesthetic value to their normative status. The representational richness made possible by recycling our own representations to represent other people's representations, or to represent other attributes of representations, is our most distinctive trait— one of these amazingly brilliant solutions that natural selection stumbles upon. However, whereas it is much simpler to rely on metarepresentations than on the cumbersome solution of creating new representations from scratch every time, we still face a complex computational task. Even when we use our own representations to attribute representations to other people, a lot of work remains to be done—it cannot be metarepresentations all the way down. At some point, we need other inputs—linguistic or behavioral cues—to attribute representations. Moreover, when a representation is represented, not all the inferences that can be drawn from it are relevant. When Mary believes that John believes it's going to rain, some of the inferences that she would draw from "it's going to rain" may not be attributable to John. Maybe he doesn't mind jogging in the rain, for instance. And Mary may not draw other inferences: Maybe John will be worried because he left his book outside. Still, without a baseline—Mary's own representation— the task would jump from merely difficult to utterly intractable.

Probably more than any other cognitive trait, the ability to use our own representations to represent representations is what explains humankind's achievements. Without this skill, the complex forms of social cognition that characterize our species would have been all but impossible. It is also critical for us psychologists to understand these ideas if we want to continue our forays into human cognition.

I leave the last word to Dan Sperber, who more than any other cognitive scientist has made metarepresentations the most central explanation of our unique cognition: "Humans have the ability to represent representations. I would argue that this metarepresentational ability is as distinctive of humans, and as important in understanding their behaviour, as is echolocation for bats."*

---

* "Intuitive and reflective beliefs," *Mind and Language* 12: 1, 67-83 (1997).

WHY THE HUMAN MIND MAY SEEM TO HAVE AN ELEGANT EXPLANATION EVEN IF IT DOESN'T

NICHOLAS HUMPHREY

Emeritus professor, London School of Economics; author,
Soul Dust: The Magic of Consciousness

On reading *The Origin of Species*, Erasmus Darwin wrote to his brother Charles in 1859: "The *a priori* reasoning is so entirely satisfactory to me that if the facts won't fit in, why so much the worse for the facts." Some of the facts—such as Kelvin's calculation of the age of the Earth—looked awkard for Darwin's theory at the time. But the theory of natural selection was too beautiful to be wrong. The brother was sure the troublesome facts would have to change. And so they did.

But it doesn't always work that way. Elegance can be misleading. Consider a simple mathematical example. Given the sequence 2, 4, 6, 8, what rule would you guess is operating to generate the series? There are several theoretically possible answers. One would be the simple rule: Take the previous number, x, and compute $x + 2$. But equally valid for these data would be the much more complicated rule: Take the previous number, x, and compute

$$-\frac{1}{44} x^3 + \frac{3}{11} x^2 + \frac{34}{11}$$

For the sequence as given so far, the first rule is clearly the more elegant. And if someone, let's call her Tracey, were to maintain that since both rules work equally well she was going to make a personal choice of the second, we would surely think she was being deliberately contrarian and anti-elegant. Tracey Emin, not Michelangelo.

But suppose Tracey were to say, "I bet if we look a little further, we shall find I was right all along." And suppose, when we do look further, we find to our surprise that the next number in the sequence is not 10, but 8.91 and the next after that not 12 but 8.67. That is, the sequence we actually discover goes 2, 4, 6, 8, 8.91, 8.67. Then what had previously seemed the better rule would no longer fit the facts at all. Yet—surprise, surprise!—the second rule would still fit nicely. In this case, we should be forced to concede that Tracey's anti-elegance had won the day.

How often does the real world tease us by seeming simpler than it really is? A famous case is Francis Crick's 1957 theory of how DNA passes on instructions for protein synthesis using a "comma-free code." As Crick wrote many years later, "Naturally [we] were excited by the idea of a comma-free code. It seemed so pretty, almost elegant. You fed in the magic numbers 4 (the 4 bases) and 3 (the triplet) and out came the magic number 20, the number of the amino acids."* But alas, this lovely theory could not be squared with experimental facts. The truth was altogether less elegant.

A tease? I'm not, of course, suggesting that nature was deliberately stringing Crick along. As Einstein said, God is subtle but he is not malicious. In this case, the failure of the most elegant explanation to be the true one is presumably just a matter of bad luck. And, assuming this doesn't happen often, perhaps in general we can still expect truth and beauty to go together (as no doubt many of the other answers to this *Edge* Question will prove).

However, there is one class of cases where the elegance of an untrue theory may not be luck at all—where, indeed, complex phenomena have actually been designed to masquerade as simple

---

* Francis Crick, *What Mad Pursuit* (New York: Basic Books, 1988), 99-100.

ones, or at any rate to masquerade as such to human beings. And such cases will arise just when, in the course of evolution, it has been to the biological advantage of humans to see certain things in a particularly simple way. The designer of the pseudo-elegant explananda has been not God but natural selection.

Here is my favorite example. Individual humans appear to other humans to be controlled by the remarkable structures we call minds. But the surprising and wonderful thing is that human minds are quite easy for others to read. We've all been doing it since we were babies, using the folk theory known to psychologists as "Theory of Mind" (or sometimes as "belief desire psychology"). Theory of Mind is simple and elegant, and can be understood by a two-year-old. There's no question that it provides a highly effective way of explaining the way people behave. And this skill at mind-reading has been essential to human survival in social groups. Yet the fact is, Theory of Mind could never have worked so well unless natural selection had shaped human brains to be able to read—and to be readable by—one another in this way. Which is where the explanatory sleight-of-hand comes in. For as an explanation of how the *brain* works, Theory of Mind just doesn't add up. It's a purpose-built, oversimplified, deep, elegant myth—a myth whose inadequacy may not become apparent, perhaps, until those "extra numbers" are added by madness or by brain damage—contingencies that selection has not allowed for.

I find this explanation of the elegance of Theory of Mind beautiful.

FITNESS LANDSCAPES

STEWART BRAND
Founder, Whole Earth Catalog; *cofounder, The Well; cofounder, The Long Now Foundation; author,* Whole Earth Discipline: An Ecopragmatist Manifesto

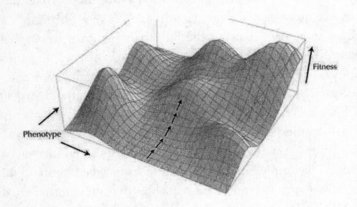

The first time I saw a fitness-landscape cartoon (in Garrett Hardin's *Nature and Man's Fate,* 1965), I knew it was giving me advice on how not to get stuck overadapted—hence overspecialized—on some local peak of fitness, when whole mountain ranges of opportunity could be glimpsed in the distance. But getting to them involved venturing "downhill" into regions of lower fitness. I learned to distrust optimality.

Fitness landscapes (sometimes called "adaptive landscapes") keep turning up when people try to figure out how evolution or innovation works in a complex world. An important critique by Marvin Minsky and Seymour Papert of early optimism about artificial intelligence warned that seemingly intelligent agents would dumbly "hill climb" to local peaks of illusory optimality and get stuck there. Complexity theorist Stuart Kauffman used fitness landscapes to visualize his ideas about the "adjacent possible" in 1993 and 2000, and that led in turn to Steven Johnson's celebration of how the "adjacent possible" works for innovation in *Where Good Ideas Come From.*

The man behind the genius of fitness landscapes was the founding theorist of population genetics, Sewall Wright (1889–1988). In 1932, he came up with the landscape as a way to visualize and explain how biological populations escape the potential trap of a local peak by imagining what might drive their evolutionary path downhill from the peak toward other possibilities. Consider these six diagrams of his:

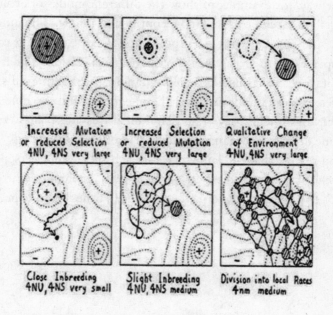

© Sewall Wright, *The Role of Mutation, Inbreeding, Crossbreeding, and Selection in Evolution,* Sixth International Congress of Genetics, Brooklyn, NY: Brooklyn Botanical Garden, 1932.

The first two illustrate how low selection pressure or a high rate of mutation (which comes with small populations) can broaden the range of a species, whereas intense selection pressure or a low mutation rate can severely limit a species to the very peak of local fitness. The third diagram shows what happens when the landscape itself shifts, and the population has to evolve to shift with it.

The bottom row explores how small populations respond to inbreeding by wandering ineffectively. The best mode of exploration Wright deemed the final diagram, showing how a species can divide into an array of races that interact with one another. That jostling crowd explores well, and it can respond to opportunity.

Fitness landscapes express so much so economically. There's no better way, for example, to show the different modes of evolution of a remote oceanic island and a continental jungle. The jungle is dense and rugged, with steep peaks and valleys, isolating countless species on their tiny peaks of high specialization. The island, with its few species, is like a rolling landscape of gentle hills with species casually wandering over them, evolving into a whole array of Darwin's finches, say. The island creatures and plants "lazily" become defenseless against invaders from the mainland.

You realize that the landscape for each species consists almost entirely of other species, all of them busy evolving right back. That's coevolution. We are all each other's fitness landscapes.

ON OCEANS AND AIRPORT SECURITY

KEVIN P. HAND

*Planetary scientist and astrobiologist; deputy chief
scientist, Solar System Exploration, NASA Jet Propulsion
Laboratory, California Institute of Technology*

It may sound odd, but much as I loathe airport security lines, I must admit that while I'm standing there, stripped down and denuded of metal, waiting to go through the doorway, part of my mind wanders to oceans that likely exist on distant worlds in our solar system.

These oceans are sheltered beneath the icy shells that cover worlds like Europa, Ganymede, and Callisto (moons of Jupiter), and Enceladus and Titan (moons of Saturn). The oceans within these worlds are liquid water, just as we know and love it here on Earth, and they have probably existed for much of the history of the solar system (about 4.6 billion years). The total volume of liquid water contained in them is at least twenty times that found on Earth. From the standpoint of our search for life beyond Earth, these oceans are prime real estate for a second origin of life and the evolution of extraterrestrial ecosystems.

But how do we know they exist? The moons are covered in ice, and we can't just look down with a spacecraft and see liquid water. That's where airport security comes into play. When you walk through an airport-security portal, you're walking through a rapidly changing magnetic field. The laws of physics dictate that if you put a conducting material in a changing magnetic field, electric currents will arise, and those electric currents will create a secondary magnetic field. This secondary field is often referred to as the induced magnetic field, because it is induced by the primary field of the portal. Within the portal are detectors that can sense when an induced field is present. When they do, the alarm goes off, and you're whisked over to the special search line.

The same fundamental physics is largely responsible for our knowledge of oceans on some of these distant worlds. Europa provides a good example. Back in the late 1990s, NASA's *Galileo* spacecraft made several flybys of Europa, and the magnetic field sensors on the spacecraft found that Europa does not have a strong internal field of its own. Instead, it has an induced magnetic field, created as a result of Jupiter's strong background magnetic field. In other words, the alarm went off.

But in order for the alarm to go off, there needed to be a conductor. And for Europa, the data indicated that the conducting layer must be near its surface. Other lines of evidence had already shown that the outer 150 kilometers or so of Europa was water, but those datasets could not help distinguish between solid ice and liquid water. For the magnetic-field data, however, ice doesn't work—it's not a good conductor. Liquid water with salts dissolved in it, like our own ocean, does work. The best fits to the data indicate that Europa has an outer ice shell about 10 kilometers thick, beneath which lies a global ocean about 100 kilometers deep. Beneath that is a rocky seafloor, which may be teeming with hydrothermal vents and bizarre otherworldly organisms.

So the next time you're in airport security and frustrated by that disorganized person in front of you who can't seem to get it through his head that his belt, wallet, and watch will all set off the alarm, just take a deep breath and think of the possibly habitable distant oceans we now know of, thanks to the same beautiful physics that's driving you nuts as you contemplate missing your plane.

PLATE TECTONICS ELEGANTLY VALIDATES CONTINENTAL DRIFT

PAUL SAFFO

Technology forecaster; managing director, foresight, Discern Analytics; consulting associate professor, Stanford University

Plate tectonics is a breathtakingly elegant explanation of a beautiful theory: continental drift. Both puzzle and answer were hiding in plain sight, right under our feet. Generations of globe-twirling schoolchildren have noticed that South America's bulge seems to fit into the gulf of Africa and that Baja California looks as if it had been cut out of the Mexican mainland. These and other more subtle clues led Alfred Wegener to propose to the German Geological Society in 1912 that the continents had once formed a single landmass. His beautiful theory was greeted with catcalls and scientific brickbats.

The problem was that Wegener's beautiful theory lacked a mechanism. Critics sneeringly pronounced that the lightweight continents could not possibly plow through a dense and unyielding oceanic crust. No one, including Wegener, could imagine a force that could cause the continents to move. It didn't help that Wegener was a meteorologist poaching in geophysical territory. He would die on an expedition to Greenland in 1930, his theory out of favor and all but forgotten.

Meanwhile, hints of a mechanism were everywhere but at once too small and too vast to be seen. Like ants crawling on a globe, puny humans missed the obvious. It would take the arrival of powerful new scientific tools to reveal the hidden forensics of continental drift. Sonar traced mysterious linear ridges running zipperlike along ocean floors. Magnetometers towed over the seabed painted symmetrical zebra-striped patterns of magnetic reversals. Earthquakes betrayed plate boundaries to listening seismographs. And radiometric dating laid out a scale reaching into deep time.

Three decades after Wegener's death, the mechanism of plate tectonics emerged with breathtaking clarity. The continents weren't plowing through anything—they were rafting atop the crust, like marshmallows stuck in a sheet of cooling chocolate. And the oceanic crust was moving like a conveyor, with new crust created in midocean spreading centers and old crust subducted, destroyed, or crumpled upward into vast mountain ranges at the boundaries where plates met.

Elegant explanations are the Kuhnian solvent that leaches the glue from old paradigms, making space for new theories to take hold. Plate tectonics became established beyond a doubt in the mid-1960s. Contradictions suddenly made sense, and ends so loose no one thought they were remotely connected came together. Continents were seen for the wanderers they were. The Himalayas were recognized as the result of a pushy Indian plate smashing into its Eurasian neighbor, and it became obvious that an ocean was being born in Africa's Great Rift Valley. Mysteries fell like dominoes before the predictive power of a beautiful theory and its elegant explanation. The skeptics were silenced and Wegener was posthumously vindicated.

WHY SOME SEA TURTLES MIGRATE

DANIEL C. DENNETT
Philosopher; university professor and codirector, Center for Cognitive Studies, Tufts University; author, Breaking the Spell: Religion as a Natural Phenomenon

My choice is an explanation that delights me. It may be true and may be false—I don't know, but probably somebody who reads *Edge* will be able to say, authoritatively, with suitable references. I am eager to find out. I was told some years ago that the reason that some species of sea turtles migrate all the way across the South Atlantic to lay their eggs on the east coast of South America after mating on the west coast of Africa is that when the behavior started, Gondwana was just beginning to break apart (that would be between 130 and 110 million years ago), and these turtles were just swimming across the narrow strait to lay their eggs. Each year the swim was a little longer—maybe an inch or so—but who could notice that? Eventually they were crossing the ocean to lay their eggs, having no idea, of course, why they would do such an extravagant thing.

What is delicious about this example is that it vividly illustrates several important evolutionary themes: the staggering power over millions of years of change so gradual it is essentially unnoticeable; the cluelessness of much animal behavior, even when it is adaptive; and of course the eye-opening perspective that evolution by natural selection can offer to the imagination of the curious naturalist. It also demonstrates *either* the way an evolutionary hypothesis can be roundly refuted by discoverable facts (if it is refuted) or the way it can be supported by further evidence (if in fact it is so supported).

An attractive hypothesis such as this is the beginning, not the end, of the inquiry. Critics often deride evolutionary hypotheses about prehistoric events as "just-so stories," but as a blanket condemnation this charge should be rejected out of hand. Thousands of such hypotheses—first dreamed up on slender evidence—have been tested and confirmed beyond a reasonable doubt. Thousands of others have been tested and disconfirmed. They were just-so stories until they weren't, in other words. That's the way science advances.

I have noticed that there is a pattern in the use of the "just-so story" charge: With almost no exceptions, it is applied to hypotheses about *human* evolution. Nobody seems to object that we can't know enough about the selective environment leading to whales or flowers for us to hold forth so confidently about how and why whales and flowers evolved as they did. So my rule of thumb is: If you see the "just-so story" epithet hurled, look for a political motive. You'll almost always find one. While it is no doubt true that some evolutionary psychologists have offered hypotheses about human evolution for which there is still only slender supporting evidence, and while it is also no doubt true that some evolutionary psychologists have been less than diligent in seeking further evidence to confirm or disconfirm their favorite hypotheses, this is at most a criticism of the thoroughness of some researchers in the field, not a condemnation of their method or their hypotheses. The same could be said about many other topics in evolutionary biology.

A HOT YOUNG EARTH:
UNQUESTIONABLY BEAUTIFUL AND
STUNNINGLY WRONG

CARL ZIMMER
Science writer; author, A Planet of Viruses

Around 4.567 billion years ago, a giant cloud of dust collapsed in on itself. At the center of the cloud, our sun began to burn, while the outlying dust grains began to stick together as they orbited the new star. Within a million years, those clumps of dust had become protoplanets. Within about 50 million years, our own planet had already reached about half its current size. As more protoplanets crashed into Earth, it continued to grow. All told, it may have taken another 50 million years to reach its full size, a time during which a Mars-sized planet crashed into it, leaving behind a token of its visit—our moon.

The formation of the Earth commands our greatest powers of imagination. It is primordially magnificent. But *elegant* is not the word I'd use to describe the explanation I just sketched out. Scientists did not derive it from first principles. There is no equivalent of $e = mc^2$ that predicts how the complex violence of the early solar system produced a watery planet that could support life. In fact, the only reason we now know so much about how the Earth formed is because geologists freed themselves from a seductively elegant explanation that was foisted on them 150 years ago. It was unquestionably beautiful, and stunningly wrong.

The explanation was the work of one of the greatest physicists of the 19th century, William Thomson (aka Lord Kelvin). Kelvin's accomplishments ranged from the concrete (figuring out how to lay a telegraph cable from Europe to America) to the abstract (the first and second laws of thermodynamics). Kelvin spent much of his career writing equations that could let him calculate how fast hot things got cold. He realized that he could use these equations to estimate how old the Earth is. "The mathematical theory on which these estimates are founded is very simple," Kelvin declared when he unveiled it in 1862.[*]

At the time, scientists generally agreed that the Earth had started out as a ball of molten rock and had been cooling ever since. Such a birth would explain why rocks are hot at the bottom of mine shafts: The surface of the Earth was the first part to cool, and ever since, the remaining heat inside the planet had been flowing out into space. Kelvin reasoned that over time the planet should steadily grow cooler. He used his equations to calculate how long it should take for a molten sphere of rock to cool to Earth's current temperature, with its observed rate of heat flow. His verdict was a brief 98 million years.

Geologists howled in protest. They didn't know how old the Earth was, but they thought in billions of years, not millions. Charles Darwin—who was a geologist first and a biologist later—estimated that it had taken 300 million years for a valley in England to erode into its current shape. The Earth itself, Darwin argued, was far older. And when he published his theory of evolution he took it for granted that the Earth was inconceivably old; the luxury of time provided room for evolution to work slowly and imperceptibly.

Kelvin didn't care. His explanation was so elegant, so beautiful, so simple, that it had to be right. It didn't matter how much trouble it caused for other scientists who were ignoring thermodynamics.

---

[*] "On the Secular Cooling of the Earth," *Trans. Roy. Soc. Edinburgh* XXIII, 167-9 (1864). Read April 28, 1862.

In fact, Kelvin made even more trouble for geologists when he took another look at his equations. He decided his first estimate had been too generous. The Earth might be only 10 million years old.

It turned out that Kelvin was wrong, but not because his equations were ugly or inelegant. They were flawless. The problem lay in the model of the Earth to which Kelvin applied his equations.

The story of Kelvin's refutation got a bit garbled in later years. Many people (myself included) have mistakenly claimed that his error stemmed from his ignorance of radioactivity. Radioactivity was not discovered until the early 1900s, as physicists worked out quantum physics. The physicist Ernest Rutherford declared that the heat released as radioactive atoms broke down inside the Earth kept it warmer than it would be otherwise. Thus a hot Earth did not have to be a young Earth.

It's true that radioactivity does give off heat, but there isn't enough inside the planet to account for the heat flowing out of it. Instead, Kelvin's real mistake was assuming that the Earth was just a solid ball of rock. In reality, the rock flows like syrup, its heat lifting it up toward the crust, where it cools and then sinks back into the depths once more. This stirring of the Earth is what causes earthquakes, drives old crust down into the depths of the planet, and creates fresh crust at ocean ridges. It also drives heat up into the crust at a much greater rate than Kelvin envisioned.

That's not to say that radioactivity didn't have its part to play in showing that Kelvin was wrong. Physicists realized that the ticktock of radioactive decay created a clock they could use to estimate the age of rocks with exquisite precision. Thus we can now say that the Earth is not just billions of years old but 4.567 billion.

Elegance unquestionably plays a big part in the advancement of science. The mathematical simplicity of quantum physics is lovely to behold. But in the hands of geologists, quantum physics has brought to light the glorious, messy, and very inelegant history of our planet.

SEXUAL-CONFLICT THEORY

DAVID M. BUSS

Professor of psychology, University of Texas, Austin; coauthor (with Cindy M. Meston), Why Women Have Sex

A fascinating parallel has occurred in the traditionally separate disciplines of evolutionary biology and psychology. Biologists historically viewed reproduction as an inherently cooperative venture. A male and female would couple for the shared goal of reproduction of mutual offspring. In psychology, romantic harmony was presumed to be the normal state. Major conflicts within romantic couples were, and still are, typically seen as signs of dysfunction. A radical reformulation, embodied by sexual-conflict theory, changes these views.

Sexual conflict occurs whenever the reproductive interests of an individual male and individual female diverge—or, more precisely, when the "interests" of their genes diverge. Sexual-conflict theory defines the many circumstances in which discord is predictable and entirely expected.

Consider deception on the mating market. If a man is pursuing a short-term mating strategy and the woman in whom he is sexually interested is pursuing a long-term mating strategy, conflict between them is virtually inevitable. Men are known to feign long-term commitment, interest, or emotional involvement for the goal of casual sex, interfering with women's long-term mating strategy. Men have evolved sophisticated strategies of sexual exploitation; conversely, women sometimes present themselves as costless sexual opportunities and then invade a man's mating mind so successfully that he wakes up one morning and realizes he can't live without her—one version of the bait-and-switch tactics in women's evolved arsenal.

Once coupled in a long-term romantic union, a man and a woman often still diverge in their evolutionary interests. Sexual infidelity on the part of the woman might benefit her by securing superior genes for her progeny, an event with catastrophic costs to her hapless partner, who unknowingly devotes resources to a rival's child. From a woman's perspective, a man's infidelity risks diversion of precious resources to rival women and their children and poses the danger of losing the man's commitment entirely. Sexual infidelity, emotional infidelity, and resource infidelity are such common sources of sexual conflict that theorists have coined distinct phrases for each.

But all is not lost. As evolutionist Helena Cronin has eloquently noted, sexual conflict arises in the context of sexual cooperation. The following evolutionary conditions for sexual cooperation are well specified: when relationships are entirely monogamous; when there is zero probability of infidelity or defection; when the couple produces offspring, the shared vehicles of their genetic cargo; and when joint resources cannot be differentially channeled, such as to one set of in-laws versus another. These conditions are sometimes met, leading to great love and harmony between a man and a woman.

The prevalence of deception, sexual coercion, stalking, intimate-partner violence, murder, and the many forms of infidelity reveal that conflict between the sexes is ubiquitous. Sexual-conflict theory, a logical consequence of modern evolutionary genetics, provides the most beautiful theoretical explanation for these darker sides of human sexual interaction.

THE SEEDS OF HISTORICAL DOMINANCE

DAVID PIZARRO

Associate professor of psychology, Cornell University

One of the most elegant explanations I have ever encountered in the social sciences comes courtesy of Jared Diamond, outlined in his wonderful book *Guns, Germs, and Steel.* Diamond attempts to answer an enormously complex and historically controversial question—why certain societies achieved historical dominance over others—by appealing to a set of basic differences in the physical environments from which these societies emerged, such as differences in the availability of plants and animals suitable for domestication.

These differences, Diamond argues, gave rise to a number of specific advantages, such as greater immunity to disease, that were directly responsible for the historical success of some societies. I'm not an expert in this domain, so I realize that Diamond's explanation might well be misguided—yet the appeal to such basic mechanisms in order to explain such a wide set of complex observations is so deeply satisfying that I hope he's right.

THE IMPORTANCE OF INDIVIDUALS

HOWARD GARDNER
Hobbs Professor of Cognition and Education, Harvard Graduate
School of Education; author, Truth, Beauty, and Goodness
Reframed: Educating for the Virtues in the 21st Century

I consider myself a scientist, and the theory of evolution is central to my thinking. I am a social scientist and have been informed by insights from many social sciences, including economics. Yet I have little sympathy with hegemonic attempts to explain all human behaviors via evolutionary psychology, via rational-choice economics, and/or by a combination of these two frameworks.

In a planet occupied now by nearly 7 billion inhabitants, I am amazed by the difference one human being can make. Think of classical music without Mozart or Stravinsky; of painting without Caravaggio, Picasso, or Pollock; of drama without Shakespeare or Beckett. Think of the incredible contributions of Michelangelo or Leonardo, or, in recent times, the outpouring of deep feeling at the death of Steve Jobs (or, for that matter, Michael Jackson or Princess Diana). Think of human values in the absence of Moses or Christ.

Alas, not all singular individuals make a positive difference. The history of the 20th century would be far happier had it not been for Hitler, Stalin, or Mao (or the 21st century without Bin Laden). But in reaction to these individuals, there sometimes arise more praiseworthy figures: Konrad Adenauer in Germany, Mikhail Gorbachev in the Soviet Union, Deng Xiaoping in China. These successors also make a signal difference.

I consider Mahatma Gandhi to be the most important human being of the last millennium. His achievements in India speak for themselves. But even if Gandhi had not contributed vital energy and leadership to his own country, he had enormous influence on peaceful resisters across the globe: Nelson Mandela in South Africa, Martin Luther King Jr. in the United States, and the solitary figures in Tiananmen Square in 1989 and Tahrir Square in 2011.

Despite the laudatory efforts of scientists to ferret out patterns in human behavior, I continue to be struck by the impact of single individuals, or of small groups, working against the odds. As scholars, we cannot and should not sweep these instances under the investigative rug. We should bear in mind anthropologist Margaret Mead's famous injunction: "Never doubt that a small group of thoughtful committed citizens can change the world; indeed, it is the only thing that ever has."

SUBJECTIVE ENVIRONMENT

ANDRIAN KREYE
Editor, The Feuilleton *(arts and essays) of the German
daily* Süddeutsche Zeitung, *Munich*

Explanations tend to be at their most elegant when science distills the meanderings of philosophy into fact. I was looking for explanations for an observation when I came across the theory of *Umwelt* vs. *Umfeld* (loosely, "environment" vs. "surroundings") by the Estonian biologist and founding father of biosemiotics Jakob von Uexküll. According to his definition, *Umwelt* is the subjective environment, as perceived and acted on by an organism, whereas *Umfeld* is the objective environment, which encompasses and acts on all organisms within it.

My observation had been a mere notion of the major difference between my native Europe and America, my adopted continent for a couple of decades. In Europe, the present is perceived as the endpoint of history. In America, the present is perceived as the beginning of the future. Philosophy or history, I hoped, would have an explanation for such a fundamental yet simple difference. Both can deliver parts of an explanation, of course. The different paths the histories of ideas and the histories of the countries have taken just in the past 200 years are astounding.

Uexküll's definition of the subjective environment as published in his book *Umwelt und Innnenwelt der Tiere* (*Environments and Inner Worlds of Animals*, published in 1909 in the language of his German exile) puts both philosophy and history into perspective and context. Distrusting theories, he wanted ideas to persist in nature, putting his notion of the subjective environment to the test in the Indian Ocean, the Atlantic, and the Mediterranean.

He observed simple creatures like sea anemones, sea urchins, and crustaceans, but the famous example illustrating his theory was the tick. Here he found a creature whose perception and actions could be defined by three parameters. Ticks perceive their surroundings by the directions of up and down, by warm and cold, and by the presence or absence of butyric acid. Their actions to survive and procreate are crawling, waiting, and gripping.

This model led him to define not just environment but also time as a subjective notion. He found any organism's perception of time as subjective as its perception of space, and defined by the very perceptions and actions that create the organism's subjective environment.

If subjective time is defined by the experiences and actions of an organism, the context of a continent's history, with its myriads of parameters, turns philosophy and history into mere factors in a complex environment of collective perception. Now, here was an elegant explanation for a rather simple observation. Making it even more elegant is the notion that in the context of a continent's evolution, such factors as geography, climate, food, and culture will figure in the perception of both the subjective environment and the subjective time, making it impossible to prove or disprove the explanation scientifically. Having rendered philosophy to just one of many parameters, it thus reduces its efforts to discredit Jakob von Uexküll's definition of the subjective environment to mere meanderings.

MY FAVORITE ANNOYING ELEGANT EXPLANATION: QUANTUM THEORY

RAPHAEL BOUSSO

Professor of theoretical physics, University of California–Berkeley

My favorite elegant explanations will already have been picked by others who turned in their homework early. Although I am a theoretical physicist, my choice could easily be Darwin. Closer to my area of expertise there is general relativity: Einstein's realization that free fall is a property of spacetime itself, which readily resolved a great mystery (why gravity acts in the same way on all bodies). So, in the interest of diversity, I will add a modifier and discuss my favorite *annoying* elegant explanation: quantum theory.

As explanations go, few are broader in applicability than the revolutionary framework of quantum mechanics, which was assembled in the first quarter of the 20th century. Why are atoms stable? Why do hot things glow? Why can I move my hand through air but not through a wall? What powers the sun? The strange workings of quantum mechanics are at the core of our remarkably precise and quantitative understanding of these and many other phenomena.

And strange they certainly are. An electron takes all paths between the two points at which it is observed, and it is meaningless to ask which path it actually took. We must accept that its momentum and position cannot both be known with arbitrary precision. For a while, we were even expected to believe that there are two different laws for time evolution: Schrödinger's equation governs unobserved systems, but the mysterious "collapse of the wave function" kicks in when a measurement is performed.

The latter, with its unsettling implication that conscious observers might play a role in fundamental theory, has been supplanted, belatedly, by the notion of decoherence. The air and light in a room, which in classical theory would have little effect on a measuring apparatus, fundamentally alter the quantum-mechanical description of any object that is not carefully insulated from its environment. This, too, is strange. But do the calculation and you will find that what we used to call wave-function collapse need not be postulated as a separate phenomenon. Rather, it emerges from Schrödinger's equation, once we take the role of the environment into account.

Just because quantum mechanics is strange doesn't mean it's wrong. The arbiter is nature, and experiments have confirmed many of the most bizarre properties of this theory. Nor does quantum mechanics lack elegance: It's a rather simple framework with enormous explanatory power. What annoys me is this: *We do not know for sure that quantum mechanics is wrong.*

Many great theories in physics carry within them a seed of their demise. This seed is a beautiful thing. It hints at profound discoveries and conceptual revolutions still to come. One day, the beautiful explanation that has just transformed our view of the universe will be supplanted by another, even deeper insight. Quantitatively, the new theory must reproduce all the experimental successes of the old one. But qualitatively, it is likely to rest on novel concepts, allowing for hitherto unimaginable questions to be asked and knowledge to be gained.

Newton, for instance, was troubled by the fact that his theory of gravitation allowed for instant communication across arbitrarily large distances. Einstein's general theory of relativity fixed this problem, and as a by-product gave us dynamical spacetime, black holes, and an expanding universe that probably had a beginning.

General relativity, in turn, is only a classical theory. It rests on a demonstrably false premise: that position and momentum can be known simultaneously. This may be a good approximation for apples, planets, and galaxies—large objects, for which gravitational interactions tend to be much more important than they are for the tiny particles of the quantum world. But as a matter of principle, the theory is wrong. The seed is there. General relativity cannot be the final word; it can only be an approximation to a more general quantum theory of gravity.

But what about quantum mechanics itself? Where is its seed of destruction? Amazingly, it's not obvious that there is one. The very name of the great quest of theoretical physics—quantizing general relativity—betrays an expectation that quantum theory will remain untouched by the unification we seek. String theory—in my view, by far the most successful, if incomplete, result of this quest—is strictly quantum mechanical, with no modifications whatsoever to the framework that was completed by Heisenberg, Schrödinger, and Dirac. In fact, the mathematical rigidity of quantum mechanics makes it difficult to conceive of any modifications, whether or not they are called for by observation.

Yet there are subtle hints that quantum mechanics, too, will suffer the fate of its predecessors. The most intriguing, in my mind, is the role of time. In quantum mechanics, time is an essential evolution parameter. But in general relativity, time is just one aspect of spacetime, a concept that we know breaks down at the singularity deep inside a black hole. Where time no longer makes sense, it is hard to see how quantum mechanics could still reign. As quantum mechanics surely spells trouble for general relativity, the existence of singularities suggests that general relativity may also spell trouble for quantum mechanics. It will be fascinating to watch this battle play out.

EINSTEIN'S REVENGE: THE NEW GEOMETRIC QUANTUM

ERIC R. WEINSTEIN
Mathematician and economist; principal, Natron Group

The modern theory of the quantum has only recently come to be understood to be far more exquisitely geometric than Einstein's general relativity. How this came to be discovered over the last forty years is a fascinating story that has, to the best of my knowledge, never been fully told, as it is not particularly popular with the people who created this stunning achievement.

The story begins at some point around 1973–74, when our consensus picture of fundamental particle theory stopped advancing. This stasis, known as the Standard Model of Particle Physics, seemed initially little more than a temporary resting place on the relentless path toward progress in fundamental physics, and theorists wasted no time proposing new theories in the expectation that they would be quickly confirmed by experimentalists looking for novel phenomena. But that expected entry into the promised land of new physics turned into a forty-year period of half-mad tribal wandering in an arid desert, all but devoid of new phenomena.

Yet just as particle theory was failing to advance in the mid-1970s, something amazing was quietly happening over lunch at the State University of New York at Stony Brook. There, Nobel physics laureate C. N. Yang and geometer (and soon to be billionaire) Jim Simons had started an informal seminar to understand what, if anything, modern geometry had to do with quantum field theory. Their shocking discovery was that both geometers and quantum theorists had independently got hold of different collections of insights into a common structure that each group had discovered for themselves. A Rosetta stone of sorts called the Wu-Yang dictionary was quickly assembled by the physicists, and Isadore Singer of MIT took these results from Stony Brook to his collaborator Michael Atiyah in Oxford, where their research with Nigel Hitchin began a geometric renaissance in physics-inspired geometry that continues to this day.

While the Stony Brook story is less discussed by today's younger mathematicians and physicists, it is not a point of contention between the various members of the community. The controversial part of this story, however, is that a hoped-for golden era of theoretical physics did not materialize or produce a new consensus theory of elementary particles. Instead the interaction highlighted the strange idea that, just possibly, quantum theory was actually a natural and elegant body of pure geometry that had fallen into an abysmal state of dilapidation putting it beyond mathematical recognition. By this reasoning, the mathematical train wreck of modern quantum field theory was able to cling to life by its fingernails and survive numerous near-death experiences, confronting mathematical rigor only because it was being held together by a natural infinite-dimensional geometry which is, to this day, only partially understood.

In short, most physicists were trying and failing to quantize Einstein's geometric theory of gravity because they were first meant to go in the opposite and less glamorous direction of geometrizing the quantum instead. Unfortunately for physics, mathematicians had dropped the ball and not sufficiently developed the geometry of infinite-dimensional systems (such as the Standard Model), which would have been analogous to the four-dimensional Riemannian geometry appropriated from mathematics by Einstein.

This reversal could well be thought of as Einstein's revenge on the excesses of quantum triumphalism, served ice-cold decades after his death: The more researchers dreamed of becoming the Nobel-winning physicists to quantize geometric gravity, the more they were rewarded only as mathematicians for the relatively remedial task of geometrizing the quantum. The more they claimed that the "power and glory" of string theory (a failed piece of 1970s subatomic physics which has mysteriously lingered into the 21st century) was the "only game in town," the more it looked like strings-based unification, lacking testable predictions, was itself sinking with a *glug* to the bottom of the sea.

What we learned from this episode was profound. If the physicists failed, it was only in their own terms that they went down to defeat. Just as in an earlier era, in which a number of physicists retooled to become the first generation of molecular biologists, physicists came to dominate much of modern geometry in the last four decades, scoring numerous successes that will stand the tests of time. Likewise their quest to quantize geometry backfired in the most romantic and elegant way possible, by instead geometrizing the quantum which, in hindsight, was needed to fill in a gaping hole left by the mathematical geometers. This lacuna would have been discovered sooner or later by mathematicians, as it is by now seen as an entirely natural piece of pure mathematics. Quantum field theory, despite its name, turns out really to be a piece of pure mathematics developed by ingenious amateurs out of the need to unpack the consequences of the fundamental equations representing the true physical content.

But the most important lesson is that, at a minimum, Einstein's minor dream has already been realized as something of a group effort. All known physical phenomena can now be recognized as fashioned from the pure marble of geometry, through the efforts of a pantheon of new giants with less familiar names, like Quillen, Singer, Simons, Atiyah, Witten, Penrose, Yang, Schwartz, Seiberg, Segal, Hitchin, and Jackiw. This explains, in advance of unification, that the source code of the universe is likely to be a purely geometric operating system written in a single programming language. While that leaves the quest for the unifying physics unfinished and the marble something of a motley patchwork of colors, it suggests that the leaders during the years of the Standard Model have put this period of stasis to good use for the benefit of those of us who hope to follow.

WHAT TIME IS IT?

DAVE WINER
Software developer; founder, UserLand Software;
editor, Scripting News *weblog*

A few years ago, I heard it said that only old-fashioned folk wear watches. But I thought I would always wear a watch. Today I don't wear a watch.

How do I find the time? Either I do without or I keep my eyes fixed on a screen that has the time in the upper-right corner. It's gotten so that I resent that reality doesn't display the time in the upper-right corner.

REALISM AND OTHER METAPHYSICAL HALF-TRUTHS

TANIA LOMBROZO

Assistant professor of psychology, University of California–Berkeley

The deepest, most elegant, and most beautiful explanations are the ones we find so overwhelmingly compelling that we don't even realize they're there. It can take years of philosophical training to recognize their presence and evaluate their merits. Consider the following three examples:

Realism.

We explain the success of our scientific theories by an appeal to what philosophers call realism—the idea that they are more or less true. In other words, chemistry "works" because atoms actually exist, and hand-washing prevents disease because there really are loitering pathogens.

Other minds.

We explain why people act the way they do by positing that they have minds more or less like our own. We assume they have feelings, beliefs, and desires, and that they are not (for instance) zombie automata that convincingly act as if they have minds. This requires an intuitive leap.

Causation.

We explain the predictable relationship between some events we call causes and others we call effects by an appeal to a mysterious power called causation. Yet, as noted by 18th-century philosopher

David Hume, we never "discover anything but one event following another," and we never directly observe "a force or power by which the cause operates, or any connection between it and its supposed effect."*

These explanations are at the core of humans' understanding of the world—of our intuitive metaphysics. They also illustrate the hallmarks of a satisfying explanation: They unify many disparate phenomena by appealing to a small number of core principles. In other words, they are broad but simple. Realism can explain the success of chemistry, but also of physics, zoology, and deep-sea ecology. A belief in other minds can help someone understand politics, their family, and *Middlemarch*. And assuming a world governed by orderly, causal relationships helps explain the predictable associations between the moon and the tides as well as that between caffeine consumption and sleeplessness.

Nonetheless, each explanation has been seriously attacked at one point or another. Take realism, for example. While many of our current scientific theories are admittedly impressive, they come at the end of a long succession of failures: *Every* past theory has been wrong. Ptolemy's astronomy had a good run, but then came the Copernican revolution. Newtonian mechanics is truly impressive, but it was ultimately superseded by contemporary physics. Modesty and common sense suggest that, like their predecessors, our current theories will eventually be overturned. But if they aren't true, why are they so effective? Intuitive realism is at best a metaphysical half-truth, albeit a fairly harmless one.

From these examples I draw two important lessons. First, the depth, elegance, and beauty of our intuitive metaphysical explanations can be a liability. These explanations are so broad and so

---

* *Enquiry Concerning Human Understanding*, Section 7: The Idea of Necessary Connection

simple that we let them operate in the background, constantly invoked but rarely scrutinized. As a result, most of us can't defend them and don't revise them. Metaphysical half-truths find a safe and happy home in most human minds.

Second, the depth, elegance, and beauty of our intuitive metaphysical explanations can make us appreciate them *less* rather than more. Like a constant hum, we forget that they are there. It follows that the explanations most often celebrated for their virtues—explanations such as natural selection and relativity—are importantly different from those that form the bedrock of intuitive beliefs. Celebrated explanations have the characteristics of the solution to a good murder mystery. Where intuitive metaphysical explanations are easy to generate but hard to evaluate, scientific superstars like evolution are typically the reverse: hard to generate but easy to evaluate. We need philosophers like Hume to nudge us from complacency in the first case and scientists like Darwin to advance science in the second.

ALL WE NEED IS HELP

SEIRIAN SUMNER

Research fellow in the evolution of sociality, Institute of Zoology, Zoological Society of London

I play this game with my kids. It's a "Guess who?" game: Think of an animal, person, or object, and then try to describe it to another person without giving away the identity. The other person has to guess what/who you are. You have to get in character and tell a story: What do you do, how do you feel, what do you think and want?

Let's have a go. Read the character scenarios below and see if you can guess who/what they are.

"It's just not fair! Mum says I'm getting in the way, I'm a layabout, and she can't afford for me to stay with her anymore. But I like being in a big family, and I don't want to leave. Why take the risk of leaving home? Who knows what lies out there! Mum says that if I am to stay home, we'd need some kind of 'glue' to keep us from drifting apart. Now, glue is costly, and she says she hasn't the time or energy to make it, since she's busy making babies. But then I had this brilliant idea: How about *I* make the glue, using a bit of cell wall (Mum won't mind), add some glycoproteins (they're a bit sticky, so I have to promise Mum I'll wash my hands afterward), and *bingo*! Job done: We've got ourselves a nice cosy extracellular matrix! I'm happy doing the bulk of the work, so long as Mum keeps giving me more siblings. I suggested this to Mum last night, and guess what? She said yes! But she also said I'm out the door if I don't keep up my side of the bargain. No free-riders. . . ."

Who am I? I am a unicell becoming multicellular. If I group with my relatives, then someone needs to pay the cost of keeping us together—the extracellular matrix. I don't mind paying that cost if I benefit from the replication of my own genes through my relatives.

OK, that's a tough one. Try this one:

"I'm probably what you'd call the maternal type. I like having babies, and I've probably already had too many this year—at least that's what my children tell me. But I seem to be pretty good at it. I love them all equally, obviously. Damn hard work, though, especially since their father didn't stick around. Interested in only one thing, he was, and off in a flash. But I can't see my latest babies surviving unless I get some help around the place—all these mouths to feed, no time to clear up the place. So I said to my oldest the other day, 'How's about it, kid? Fancy helping your old ma out around the place a bit? Here's the deal: You go find some food while I just squeeze out a few more siblings for you. Remember, kid, I'm doing this for you—all these siblings I'm making, it'll pay off in the long run. One day, one of them will be a ma just like me—imagine that! And you'll still be reaping the benefits from her, too, long after you and I are gone. This way you don't ever have to worry about sex, men, or any of that sperm stuff. Your old ma's got everything you need, right here. All you have to do is feed us and clear out the mess. There's a good kid—off you go, but don't talk to strangers, especially men!'"

Who am I? I am an insect becoming a society. If I nest alone, I have to find food, which means leaving my young unprotected. If some of my grown-up children stay home and help me, they can go out foraging, leaving me to defend the young. I can have even more babies this way, which my children love, as this means more and more of their genes are passed on through their siblings. Anyway, it's a pretty tough world out there right now for youngsters. It's much less risky to stay at home.

A little tweaking of the details in the above sketches and I could equally be a gene becoming a genome, or a prokaryote becoming a eukaryote. I am part of the same fundamental event in evolution's playground. I am the evolution of helping and cooperation. I am the major transition that shapes all levels of biological complexity. The reason I happen is because I help others like me, and we agree on a division of labor (OK, there are some fights, but we balance conflict with cooperation, and sometimes a little coercion doesn't go amiss). And the reason I help is not because it makes me feel good but because, paradoxically, I benefit from helping. My secret? I'm pretty selective: I like to help my relatives, because they end up also helping me, by passing on our shared genes. I've embraced the transition from autonomy to cooperation, and it feels good!

The evolution of cooperation and helping behavior is a beautiful and simple explanation for how nature got complex, diverse, and wonderful. It's not restricted to the charismatic meerkats or the fluffy bumblebees. It's a general phenomenon that sweeps across the good, bad, and ugly sides of nature, generating the biological hierarchies that characterize the natural world. Groups of individuals (genes, prokaryotes, single-celled and multicellular organisms) that could previously replicate independently come together to form new, more complex individuals in their own right. This new collective individual can replicate only as a whole. Take each component in isolation and it is unable to function or pass genes on to the next generation.

The simplest, most elegant rule in the natural world explains why this complexity evolves: William Hamilton's 1964 inclusive-fitness theory, which encompasses the essence of natural selection. Entities cooperate because it increases their fitness—their chance of passing on genes to the next generation. Receivers of help benefit from enhanced personal reproduction—direct fitness. Helpers benefit from the propagation of the genes they share with the relatives they help—indirect fitness. We still have solitary insects, single-celled organisms, and prokaryotes, because the conditions need to be right for division of labor to evolve: The benefits must outweigh the costs, and this sum is affected by the options available to independent replicating entities. Ecology and environment play a role, as well as kinship. The resulting division of labor is the fundamental basis of societal living, uniting genes into genomes, uniting mitochondria and prokaryotes to produce eukaryotes, uniting unicellular organisms into multicellular ones and solitary animals into eusocieties. Without the evolution of helping and division of labor, there would be no eukaryotes, no multicellular organisms, no animal societies—in short, our planet would be barren and dull.

We have understood this simple concept for almost half a century now. It's only very recently, however, that we have realized that the evolution of helping explains not just the transition to eusociality in insects (for which Hamilton originally developed the theory) but also the evolution of major transitions to biological complexity in general. Among others, Andrew Bourke produced an insightful synthesis of this unified framework for the origins of biological complexity in his recent book *Principles of Social Evolution*. This satisfyingly simple explanation makes the complexities of the world less mysterious but no less wonderful.

If only adults played children's games more often, perhaps we'd stumble across other simple explanations for the complexities of life.

IN THE BEGINNING IS THE THEORY

HELENA CRONIN

*Codirector of the London School of Economics' Centre for Philosophy
of Natural and Social Science; author,* The Ant and the Peacock:
Altruism and Sexual Selection from Darwin to Today

Let's eavesdrop on an exchange between Charles Darwin and Karl
Popper. Darwin, exasperated at the crass philosophy of science ped-
dled by his critics, exclaims, "How odd it is that anyone should not
see that all observation must be for or against some view if it is to
be of any service!"[*] And when the conversation turns to evolution,
Popper observes, "All life is problem solving," noting that "from the
amoeba to Einstein, the growth of knowledge is always the same."[†]

There is a confluence in their thinking. Though traveling by
different pathways, they arrived at the same insight. It has to do
with the primacy and fundamental role of theories—of ideas,
hypotheses, perspectives, views, dispositions, and the like—in the
acquisition and growth of knowledge. Darwin was right to stress
that such primacy is needed "if [the observation] is to be of any
service." But the role of a "view" also goes far deeper. As Darwin
knew, it is impossible to observe at all without some kind of view.
If you are unconvinced, try this demonstration, one that Popper
liked to use in lectures. "Observe!" Have you managed that? No.
Because, of course, you need to know, "Observe what?" All observation is in the light of some theory; all observation must be in

---

[*] Darwin to Henry Fawcett, 18 Sept. 1861, Darwin Correspondence Database,
    http://www.darwinproject.ac.uk/entry-3257.

[†] *Objective Knowledge: An Evolutionary Approach* (Oxford, UK: Oxford University
    Press, 1972), 261.

the light of some theory. So all observation is theory-laden—not sometimes, not contingently, but always and necessarily.

This is not to depreciate observation, data, facts. On the contrary, it gives them their proper due. Only in the light of a theory, a problem, the quest for a solution, can they speak to us in revealing ways.

Thus the insight is immensely simple. But it has wide relevance and great potency. Hence its elegance and beauty.

Here are two examples, first from Darwin's realm, then from Popper's.

- Consider the tedious but tenacious argument: "genes vs. environment." I'll take a well-studied case. Indigo buntings migrate annually over long distances. To solve the problem of navigation, natural selection equipped them with the ability to construct a mental compass by studying the stars in the night sky, boy-scout fashion, during their first few months of life. The fount of this spectacular adaptation is a rich source of information that natural selection, over evolutionary time, has packed into the birds' genes—in particular, information about the rotation of the stellar constellations. Thus buntings that migrate today can use the same instincts and the same environmental regularities to fashion the same precision-built instrument as did their long-dead ancestors.

And all adaptations work in this way. By providing the organism with innate information about the world, they open up resources for the organism to meet its own distinctive adaptive needs; thus natural selection creates the organism's own tailor-made, species-specific environment. And different adaptive problems therefore give rise to different environments; so different species, for example, have different environments.

Thus what constitutes an environment depends on the organism's adaptations. Without innate information, carried by the genes, specifying what constitutes an environment, no environments would exist. And thus environments, far from being separate from biology, autonomous and independent, are themselves in part fashioned by biology. Environment is therefore a biological issue, an issue that necessarily begins with biologically stored information.

But aren't we anyway all interactionists now—no longer genes vs. environment but gene/environment interaction? Yes, of course; interaction is what natural selection designed genes to do. Bunting genes are freighted with information about how to learn from stars because stars are as vital a part of a bunting's environment as is the egg in which it develops or the water it drinks. Buntings without stars are destined to be buntings without descendants. But interaction is not parity; the information must come first. Just try this parity test. Try specifying "an" environment without first specifying whether it is the environment of a bunting or a human, a male or a female, an adaptation for bird navigation or for human language. The task is impossible; the specification must start from the information stored in adaptations. And here's another challenge to parity. Genes use environments for a purpose: self-replication. Environments, however, have no purposes; so they do not use genes. Thus bunting genes are machines for converting stars into more bunting genes; but stars are not machines for converting bunting genes into more stars.

- The second example has to do with the notion of objectivity in science. Listen further to Darwin's complaint about misunderstandings over scientific observation: "How profoundly ignorant [this critic] must be of the very soul of observation! About thirty years ago, there was much talk that geologists ought only to observe and not theorise; and I well remember some one saying that at this rate a man might as well go into a gravel-pit and count the pebbles and describe the colours.""

One hundred and fifty years later, variants of that thinking still stalk science. Consider the laudable but now somewhat tarnished initiative to establish evidence-based policymaking. What went wrong? All too often, objective evidence was taken to be data uncontaminated by the bias of a prior theory. But without "the very soul" of a theory as guidance, what constitutes evidence? Objectivity isn't to do with stripping out all presuppositions. Indeed, the more that's considered possible or desirable, the greater the undetected, uncriticized presuppositions and the less the objectivity. At worst, a desired but unstated goal can be smuggled in at the outset. And the upshot? This well-meant approach is often justifiably derided as "policy-based evidence-making."

An egregious example from my own recent experience, which still has me reeling with dismay, was from a researcher on "gender diversity" whose concern was discrimination against women in the professions. He claimed that his research was absolutely free of any prior assumptions about male/female differences and that it was therefore entirely neutral and unbiased. If any patterns of differences emerged from the data, his neutral, unbiased assumption would be that they were the result of discrimination. So might he accept that evolved sex differences exist? Yes, if it were proven. And what might such a proof look like? Here he fell silent, at a loss—unsurprisingly, given that his "neutral" hypotheses had comprehensively precluded such differences at the start. What

---

* Darwin to Fawcett, 18 Sept. 1861.

irony, that in the purported interests of scientific objectivity he ostensibly felt justified in clearing the decks of the entire wealth of current scientific findings.

The Darwin-Popper insight, in spite of its beauty, has yet to attract the admirers it deserves.

THOMPSON ON DEVELOPMENT

PAUL BLOOM
*Brooks and Suzanne Ragen Professor of Psychology & Cognitive
Science, Yale University; author,* How Pleasure Works

"Everything is the way it is because it got that way." This aphorism is
attributed to the biologist and classicist D'Arcy Thompson, and it's
an elegant summary of how Thompson sought to explain the shapes
of things, from jellyfish to sand dunes to elephant tusks. I saw this
quoted first in an *Edge* discussion by Daniel Dennett, who made the
point that this insight applies to explanation more generally—all
sciences are, to at least some extent, historical sciences.

I think it's a perfect motto for my own field of developmental
psychology. Every adult mind has two histories. There is evolution; few would doubt that some of the most elegant and persuasive explanations in psychology appeal to the constructive process
of natural selection. And there is development—how our minds
unfold over time, the processes of maturation and learning.

While evolutionary explanations work best for explaining what
humans share, development can sometimes capture how we differ.
This may be obvious: Nobody is surprised to hear that adults who
are fluent in Korean have usually been exposed to Korean when
they were children, or that adults who practice Judaism have usually been raised as Jews. But other developmental explanations are
rather interesting.

There is evidence that an adult's inability to see in stereo is due
to poor vision during a critical period in childhood. Some have
argued that the self-confidence of adult males is influenced by how
young they were when they reached puberty (because of the boost
in status caused by being bigger, even if temporarily, than their

peers). It has been claimed that smarter adults are more likely to be firstborns (because later children find themselves in environments that are, on average, less intellectually sophisticated). Creative adults are more likely to be later-borns (because they were forced to find their own distinctive niches). Romantic attachments in adults are influenced by their relationships as children with their parents. A man's pain-sensitivity is influenced by whether or not he was circumcised as a baby.

With the exception of the stereo-vision example, I don't know if any of these explanations are true. But they are elegant and non-obvious, and some of them verge on beautiful.

HOW DO YOU GET FROM A LOBSTER TO A CAT?

JOHN McWHORTER

Linguist; senior fellow, Manhattan Institute for Policy Research; author,
What Language Is (And What It Isn't and What It Could Be)

Did you ever notice that the "vein" you are told to remove from shrimp before eating them doesn't seem to ooze anything you'd be inclined to call blood? Doesn't the slime seem more like some sort of alimentary waste? That's because it is. In shrimp, you can get at the digestive system right through its back, because that's where it is. The heart's up there too, and this is the way it is in arthropods, the animal phylum that includes crustaceans and insects. Meanwhile, if you were interested in finding the shrimp's main nerve highway, you'd find it running down along its bottom side.

That seems backwards to us, because we're chordates, another big animal phylum. Chordates have the spinal nerve running down the back, with the heart and gut in front. It's as if our body plans were mirror images of arthropods', and this is a microcosm of a general split between larger classes. Arthropods are among the *protostomes*, with the guts on the back, as opposed to the *deuterostomes* that we chordates are among, with the guts up front.

Biologists have noticed this since auld lang syne, with naturalist Étienne Geoffroy Saint-Hilaire turning a dissected lobster upside down and showing that as such, its innards' arrangement resembled ours. The question was how things got this way, especially as Darwin's natural-selection theory became accepted. How could one get step-by-step from the guts on the back and the spinal cord up front to the reverse situation? More to the point, why would this be evolutionarily advantageous, which is the only reason we assume it would happen at all?

Short of imagining that the nerve cord glommed upward and took over the gut and a new gut spontaneously developed down below because it was "needed"—this idea was actually entertained for a while by one venturesome thinker—the best that biologists could do for a long time was to suppose that the arthropod plan and the chordate plan were alternative pathways of evolution from some primordial creature; just a matter of the roll of the dice, they thought.

Not only was this explanation boring, the problem was also that molecular biology was making it ever clearer that arthropods and chordates trace back to the same basic body plan in a good amount of detail. The shrimp's little segments are generated by the same genes that create our vertebral column, and so on. Which leads back to the old question: How do you get from a lobster to a cat? Biologists are converging upon an answer that combines elegance with a touch of mystery, with a scintilla of humility in the bargain.

What's increasingly thought to have happened is that some early wormlike aquatic creature with the arthropod-style body plan started swimming upside down. Creatures can do that— brine shrimp today, for example. Often it's because a creature's coloring is different on top than on the bottom, and having the top color down makes them harder for predators to see. So there would have been evolutionary advantage to such a creature turning upside down forever. In this creature, the spinal cord was up and the guts were down. By itself, this story is perhaps cute, maybe a little sad, but not much more. But suppose this little worm then evolved into today's chordates? It's hardly a stretch, given that the most primitive chordates actually are wormish, only vaguely piscine things called lancelets. And if you were moved to rip one open, you'd see that nerve cord on the back, not the front.

Molecular biology is quickly showing exactly how developing organisms can be signaled either to develop a shrimplike or a cat-like body plan along these lines. There even seems to be a missing link—there are rather vile, smelly, bottom-feeding critters called acorn worms that have nerve cords on the back and the front and guts that seem on their way to moving on down.

So the reason we humans have a backbone is not because it's somehow better to have a spinal column to break a fall backward, or anything of the sort. Roll the dice again and we could be bipeds with spinal columns running down our fronts like zippers and the guts carried in the back (this actually doesn't sound half bad). This explanation of what's called dorsoventral inversion is yet more evidence of how, under natural selection, such awesome variety can emerge in unbroken fashion from such humble beginnings. And finally it's hard not to be heartened by a scientific explanation that early adopters, like Geoffroy Saint-Hilaire, were ridiculed for espousing.

Quite often when I'm preparing shrimp, or tearing open a lobster, or contemplating what it would be like to be forced to dissect an acorn worm, or patting my cat on the belly, or giving someone a hug, I think a bit about the fact that all these bodies are built on the same plan, except that the cat's and the huggee's bodies are the legacy of some worm swimming the wrong way up in a Precambrian ocean more than 550 million years ago. It has always struck me as rather gorgeous.

GERMS CAUSE DISEASE

GREGORY COCHRAN

Consultant, Adaptive Optics; adjunct professor of anthropology,
University of Utah; coauthor (with Henry Harpending), The 10,000-
Year Explosion: How Civilization Accelerated Human Evolution

The germ theory of disease has been very successful, particularly if you care about practical payoffs like staying alive. It explains how disease can rapidly spread to large numbers of people (exponential growth), why there are so many different diseases (distinct pathogen species), and why some kind of contact (sometimes indirect) is required for disease transmission. In modern language, most disease syndromes turn out to be caused by tiny self-replicating machines whose genetic interests are not closely aligned with ours.

In fact, germ theory has been so successful that it almost seems uninteresting. Once we understood the causes of cholera and pneumonia and syphilis, we got rid of them, at least in the wealthier countries. Now we're at the point where people resist the means of victory—vaccination, for example—because they no longer remember the threat.

It's still worth studying—not just to fight the next plague but also because it has been a major factor in human history and human evolution. You can't really understand Cortez without smallpox or Keats without tuberculosis. The past is another country—don't drink the water.

It may well explain patterns we aren't even supposed to see, let alone understand. For example, human intelligence was, until recently, ineffective at addressing problems caused by microparasites, as William McNeill has pointed out in *Plagues and Peoples*. Those invisible enemies played a major role in determining human biological fitness—more so in some places than others. Consider the implications.

Lastly, when you leaf through an illustrated book on tropical diseases and gaze upon an advanced case of elephantiasis or crusted scabies, you realize that any theory that explains that much ugliness just has to be true.

DIRT IS MATTER OUT OF PLACE

CHRISTINE FINN
Archaeologist, journalist; author, Artifacts: An
Archaeologist's Year in Silicon Valley

I admire this explanation of cultural relativity, by the anthropologist Mary Douglas, for its clean lines and tidiness. I like its beautiful simplicity, the way it illuminates dark corners of misreading, how it highlights the counterconventional. Poking about in the dirt is exciting, and irreverent. It's about taking what's out of bounds and making it relevant. Douglas's explanation of "dirt" makes us question the very boundaries we're pushing.

INFORMATION IS THE RESOLUTION OF UNCERTAINTY

ANDREW LIH

Associate professor of journalism, University of Southern California; author, The Wikipedia Revolution: How a Bunch of Nobodies Created the World's Greatest Encyclopedia

Nearly everything we enjoy in the digital age hinges on this one idea, yet few people are aware of its originator or the foundations of this simple, elegant theory of information. How many know that the information age was not the creation of Bill Gates or Steve Jobs but of Claude Shannon in 1948? Shannon was a humble man and an intellectual wanderer who shunned public speaking and granting interviews. This brilliant mathematician, geneticist, and cryptanalyst formulated what would become information theory in the aftermath of World War II, when it was apparent that the war had not been just a war of steel and bullets.

If World War I was the first mechanized war, the Second World War could be considered the first struggle based around communication technologies. Unlike previous conflicts, there was heavy utilization of radio communication among military forces. This rapid remote coordination pushed the war to all corners of the globe. The field of cryptography advanced quickly, in order to keep messages secret and hidden from adversaries. Also, for the first time in combat, radar was used to detect and track aircraft, thereby surpassing conventional visual capabilities that ended on the horizon.

Claude Shannon was working on the problem of antiaircraft targeting and designing fire-control systems to work directly with radar. How could you determine the current and future position of enemy aircraft so that you could properly time artillery fire to shoot them down? The radar information about plane position was a breakthrough, but "noisy," in that it provided an approximation of location but not precisely enough to be immediately useful. After the war, this inspired Shannon and many others to think about the nature of filtering and propagating information, whether radar signals, voices (on phone calls), or video (for television). Noise was the enemy of communication, so any way to store and transmit information that rejected noise was of particular interest to Shannon's employer, Bell Laboratories, the research arm of the mid-century American telephone monopoly.

Shannon considered communication the most mathematical of the engineering sciences and turned his intellect toward this problem. Having worked on the intricacies of Vannevar Bush's differential-analyzer analog computer in his early days at MIT, and with a mathematics-heavy PhD thesis ("An Algebra for Theoretical Genetics"), Shannon was particularly well suited to understanding the fundamentals of handling information using knowledge from a variety of disciplines. By 1948 he had formed his central, simple, and powerful thesis: *Information is the resolution of uncertainty.*

As long as something can be relayed that resolves uncertainty, that is the fundamental nature of information. While this sounds obvious, it was an important point, given the different languages people speak and how one utterance can be meaningful to one person and unintelligible to another. Until Shannon's theory was formulated, it was not known how to compensate for these types of "psychological factors" appropriately. Shannon built on the work of fellow researchers Ralph Hartley and Harry Nyquist to show that coding and symbols were the key to resolving whether two communicators had a common understanding of the uncertainty being resolved.

Shannon then asked, "What is the simplest resolution of uncertainty?" To him it was the flip of the coin—heads or tails, yes or no: an event with only two outcomes. He concluded that any type of information could be encoded as a series of yes or no answers. Today we know these answers as bits of digital information, 1s and 0s, which represent everything from e-mail text, digital photos, compact-disk music, or high-definition video. That any and all information could be represented and coded in discrete bits not just approximately but perfectly, without noise or error, was a breakthrough that astonished even his peers at academic institutions and Bell Laboratories, who had despaired of inventing a simple universal theory of information.

The compact disk, the first ubiquitous digital encoding system for the average consumer, brought Shannon's legacy to the masses in 1982. It provides perfect reproduction of sound by dividing each second of musical audio waves into 44,100 slices (samples), and recording the height of each slice in digital numbers (quantization). Higher sampling rates and finer quantization raise the quality of the sound. Converting this digital stream back to audible analog sound using modern circuitry allowed for consistent high fidelity. Similar digital approaches have been used for images and video, so that today we enjoy a universe of MP3, DVDs, HDTV, and AVCHD multimedia files that can be stored, transmitted, and copied with no loss of quality.

Shannon became a professor at MIT, and over the years his students made many of the major breakthroughs of the information age, including digital modems, computer graphics, data compression, artificial intelligence, and digital wireless communication. Information theory as a novel and previously unimagined discovery has transformed nearly every aspect of our daily lives to digital, from how we work to how we live and socialize. Beautiful, elegant, and deeply powerful!

EVERYTHING IS THE WAY IT IS BECAUSE IT GOT THAT WAY

PZ MYERS

Associate professor of biology, University of Minnesota Morris; author,
Atheist Voices of Minnesota: an Anthology of Personal Stories

There's no denying that the central concept of modern biology is evolution, but I was a victim of the American public school system and I went through twelve years of education without once hearing any mention of the E word. We dissected cats, we memorized globs of taxonomy, we regurgitated extremely elementary fragments of biochemistry on exams, but we were not given any framework to make sense of it all. One reason I care very much about science education now is that mine was so poor.

The situation wasn't much better in college. There, evolution was universally assumed, but there was no remedial introduction to the topic—it was sink or swim. Determined not to drown, I sought out context—anything that would help me understand all these facts my instructors expected me to know. I found it in a used bookstore, in a book that I selected because it wasn't too thick and daunting and because when I skimmed it I could tell that it was clearly written, unlike the massive, dense, and opaque reference books my classes foisted on me. It was John Tyler Bonner's *On Development: The Biology of Form*, and it blew my mind—and also warped me permanently so that I see biology through the lens of development.

The first thing the book taught me wasn't an explanation, which was something of a relief; my classes were just full of explanations already. Bonner's book is about questions—good questions, some of which had answers and others that just hung there ripely. For instance, how is biological form defined by genetics? It's the implicit question in the title, but the book refined the questions we need to answer in order to explain the problem. Maybe that's explanation at a different level: Science isn't a body of archived facts; it's the path we follow to acquire new knowledge.

Bonner also led me to D'Arcy Wentworth Thompson and his classic book, *On Growth and Form*, which provided my favorite aphorism for a scientific view of the universe, "Everything is the way it is because it got that way." It's a subtle way of emphasizing the importance of process and history in understanding why everything is the way it is. You simply cannot grasp the concepts of science if your approach is to dissect the details in a static snapshot of its current state; your only hope is to understand the underlying mechanisms that generate that state, and how it came to be. The necessity of that understanding is implicit in developmental biology, where all we do is study the process of change in the developing embryo, but I also found it essential as well in genetics, comparative physiology, anatomy, and biochemistry. And of course it is paramount in evolutionary biology.

So my most fundamental explanation is a mode of thinking: To understand how something works, you must first understand how it got that way.

THE IDEA OF EMERGENCE

DAVID CHRISTIAN

*Professor of history at Macquarie University,
Sydney; author,* Maps of Time

One of the most beautiful and profound ideas I know, and one whose power is not widely enough appreciated, is the idea of emergence and emergent properties.

When created, our universe was pretty simple. For several hundred million years, there were no stars, hardly any atoms more complex than helium, and of course no planets, no living organisms, no people, no poetry.

Then, over 13.7 billion years, all these things appeared, one by one. Each had qualities that had never been seen before. This is creativity in its most basic and mysterious form. Galaxies and stars were the first large, complex objects, and they had strange new properties. Stars fused hydrogen atoms into helium atoms, creating vast amounts of energy and forming hot spots dotted throughout the universe. In their death throes, the largest stars created all the elements of the periodic table, while the energy they pumped into the cold space around them helped assemble these elements into utterly new forms of matter with entirely new properties. Now it was possible to form planets, bacteria, dinosaurs, and us.

Where did all these exotic new things come from? How do new things, new qualities "emerge"? Were they present in the components from which they were made? The simplest reductive arguments presume they had to be. But if so, they can be devilishly hard to find. Can you find "wateriness" in the atoms of hydrogen and oxygen that form water molecules? This is why "emergence" so often seems magical and mysterious.

But it's not, really. One of the most beautiful explanations of emergence is found in a Buddhist sutra probably composed more than 10,000 years ago, "The Questions of Milinda." (I'm paraphrasing on the basis of an online translation.)

Milinda is a great emperor. He was an actual historical figure, the Greco-Bactrian emperor Menander, who ruled a Central Asian kingdom founded by generals from Alexander the Great's army. In the sutra, Milinda meets with Nagasena, a great Buddhist sage—probably on the plains of modern Afghanistan. Milinda had summoned Nagasena because he was getting interested in Buddhism but was puzzled because the Buddha seemed to deny the reality of the self. For most of us, the sense of self is the very bedrock of reality. (When Descartes said "I think, therefore I am," he doubtless meant something like "The self is the only thing we know that exists for certain.")

So we should imagine Milinda sitting in a royal chariot, followed by a huge retinue of courtiers and soldiers, meeting Nagasena, with his retinue of Buddhist monks, for a great debate about the nature of the self, reality, and creativity. It's a splendid vision.

Milinda asks Nagasena to explain the Buddha's idea of the "self." Nagasena asks, "Sire, how did you come here?" Milinda says, "In a chariot, of course, reverend Sire."

"Sire, if you removed the wheels would it still be a chariot?"

"Yes, of course it would," says Milinda, with some irritation, wondering where this conversation is going.

"And if you removed the framework, or the flagstaff, or the yoke, or the reins, or the goadstick, would it still be a chariot?"

Eventually Milinda starts to get it. He admits that at some point his chariot would no longer be a chariot, because it would have lost the quality of chariotness and could no longer do what chariots do.

And now Nagasena cannot resist gloating, because Milinda has failed to define in what exact sense his chariot really exists. Then comes the punch line: "Your Majesty has spoken well about the chariot. It is just so with me. . . . This denomination, 'Nagasena,' is a mere name. In ultimate reality, this person cannot be apprehended."

Or, in modern language, I and all the complex things around me exist only because many things were precisely assembled. The "emergent" properties are not magical. They are really there, and eventually they may start rearranging the environments that generated them. But they don't exist "in" the bits and pieces that made them; they emerge from the arrangement of those bits and pieces in very precise ways. And that is also true of the emergent entities known as "you" and "me."

FRAMES OF REFERENCE

DIMITAR D. SASSELOV

Professor of astronomy, Harvard University; director, Harvard Origins of Life Initiative; author, The Life of Super-Earths: How the Hunt for Alien Worlds and Artificial Cells Will Revolutionize Life on Our Planet

Deep and elegant explanations relate to natural or social phenomena, and the observer often has no place in them. As a young student, I was fascinated to understand how frames of reference work—that is, to learn what it means to be an observer.

The reference-frame concept is central in physics and astronomy. For example, the study of flows relies most often on two basic frames: one in which the flow is described as it moves through space, called an Eulerian frame; and another, called a Lagrangian frame, which moves with the flow, stretching and bending as it goes. The equations of motion in the Eulerian frame seemed intuitively obvious to me, but I felt exhilarated when I understood the same flow described by equations in the Lagrangian frame.

It is beautiful. Let's think of a flow of water, a winding river. You are perched on a hill by the riverbank, observing the water flow, which is marked by a multitude of floating tree leaves. The banks of the river, the details of the surroundings, provide a natural coordinate system, just as on a geographical map; you could almost create a mental image of fixed criss-crossing lines, your frame of reference. The river's flow moves through that fixed map; you can describe the twists and turns of the currents and their changing speed, all thanks to this fixed Eulerian frame of reference, named after Leonhard Euler (1707–1783).

It turns out that you could describe the flow with equal success if, instead of standing safely atop the hill, you plunged into the river and floated downstream, observing the whirling motions of the tree leaves all around you. Your frame of reference—the one named after Joseph-Louis Lagrange (1736–1813)—is no longer fixed; instead, you're describing all motions as relative to you and to one another. Your description of the flow will match exactly the description you achieved by observing from the hill, although the mathematical equations appear unrecognizably different.

To my younger self, back then, the transformation between the two frames looked like magic. It was not deep, perhaps, but it was elegant and extremely helpful. However, it was also just the start of a journey—a journey that would pull the old frames of reference out from under me. It started with the naive picture of an unmoving Earth as the absolute frame of Aristotle, soon to be rejected and replaced by Galileo with a frame of reference in which motion is not absolute. Oh, how I loved floating with Lagrange down Euler's river, only to be unsettled again by the special relativity of Einstein and trying to comprehend the loss of simultaneity. And a loss it was.

A fundamental shift in our frame of reference, especially the one that defines our place in the world, deeply affects each and every one of us personally. We live and learn. The next generation is born into the new with no attachment to the old.

In science, it's easy. But human frames of reference go beyond mathematics, physics, and astronomy. Do we know how to transform between human frames of reference successfully? Are they more often than not "Lagrangian" and relative? Perhaps we could take a cue from science and find an elegant solution. Or at least an elegant explanation.

EPIGENETICS—THE MISSING LINK

HELEN FISHER
Biological anthropologist, Rutgers University; author, Why
Him? Why Her? How to Find and Keep Lasting Love

To me, epigenetics is the most monumental explanation to emerge
in the social and biological sciences since Darwin proposed his
theories of natural selection and sexual selection. More than 2,500
articles, many scientific meetings, the San Diego Epigenome
Center and other institutes, a five-year Epigenomics Program
launched in 2008 by the National Institutes of Health, and many
other institutions, academic forums, and people are devoted to this
new field. Although epigenetics has been defined in several ways,
all are based on the concept that environmental forces can affect
gene behavior, either turning genes on or off. As an anthropologist untrained in advanced genetics, I won't attempt to explain the
processes involved, although two basic mechanisms are known:
One involves molecules known as methyl groups that latch on to
DNA to suppress and silence gene expression; the other involves
molecules known as acetyl groups, which activate and enhance
gene expression.

The consequences of epigenetic mechanisms are likely to be
significant. Scientists hypothesize that epigenetic factors play a
role in the etiology of many diseases, conditions, and human variations, from cancers to clinical depression and mental illnesses to
human behavioral and cultural variations.

Take the Moroccan Amazighs, or Berbers, people with highly
similar genetic profiles who now reside in three different environments: Some roam the deserts as nomads, some farm the mountain slopes, some live in the towns and cities along the Moroccan

coast. And depending on where they live, up to one-third of their genes are differentially expressed, reports researcher Youssef Idaghdour.[*]

For example, among the urbanites, some genes in the respiratory system are switched on—perhaps, Idaghdour suggests, to counteract their new vulnerability to asthma and bronchitis in these smoggy surroundings. Idaghdour and his colleagues propose that epigenetic mechanisms have altered the expression of many genes in these three Berber populations, producing their population differences.

Psychiatrists, psychologists, and therapists have long been preoccupied with our childhood experiences—specifically, how these sculpt our adult attitudes and behaviors. Yet they have focused on how the brain integrates and remembers these occurrences. Epigenetic studies provide a different explanation. For example, mother rats that spend more time licking and grooming their young during the first week after birth produce infants who later become better-adjusted adults. And researcher Moshe Szyf proposes that this behavioral adjustment occurs because epigenetic mechanisms are triggered during this critical period, producing a more active version of a gene that encodes a specific protein. Then this protein, via complex pathways, sets up a feedback loop in the hippocampus of the brain—enabling these rats to cope more efficiently with stress.[†]

These behavioral modifications remain stable through adulthood. However, Szyf notes that when specific chemicals were

[*]   Y. Idaghdour, J. D. Storey, S. J. Jadallah & G. Gibson, "A Genome-Wide Gene Expression Signature of Environmental Geography in Leukocytes of Moroccan Amazighs," *PLoS Genetics* 4(4):e1000052 (2008).

[†]   I. C. G. Weaver et. al., "Epigenetic programming by maternal behavior," *Nature Neuroscience* 7, 847-54 (2004).

injected into the adult rats' brains to alter these epigenetic processes and suppress this gene expression, these well-adjusted rats became anxious and frightened. And when different chemicals were injected to trigger epigenetic processes that instead enhance the expression of this gene, fearful adult rats (rats that had received little maternal care in infancy) became more relaxed.

Genes hold the instructions; epigenetic factors direct how those instructions are carried out. And as we age, scientists report, these epigenetic processes continue to modify and build who we are. Fifty-year-old twins, for example, show three times more epigenetic modifications than do three-year-old twins; and twins reared apart show more epigenetic alterations than those who grow up together. Epigenetic investigations are proving that genes are not destiny; but neither is the environment—even in people.

Shelley Taylor has shown this. Studying an allele (genetic variant) in the serotonin system, she and colleagues showed that the symptoms of depression are visible only when this allele is expressed in combination with specific environmental conditions. Moreover, Taylor maintains that individuals growing up in unstable households are likely to suffer all their lives with depression, anxiety, specific cancers, heart disease, diabetes, or obesity.* Epigenetics at work? Probably.

Even more remarkable, some epigenetic instructions are passed from one generation to the next. Transgenerational epigenetic modifications are now documented in plants and fungi and have been suggested in mice. Genes are like the keys on a piano; epigenetic processes direct how these keys are played—modifying the tune, even passing these modifications to future generations.

---

* "Early Family Experience Can Reverse the Effects of Genes, UCLA Psychologists Report," ScienceDaily, 2006. http://www.sciencedaily.com /releases/2006/10/061012190132.htm.

Indeed, in 2010, scientists wrote in *Science* magazine that epigenetic systems are now regarded as heritable, self-perpetuating, and reversible.

If epigenetic mechanisms can not only modulate our intellectual and physical abilities but also pass these alterations on to our descendants, epigenetics has immense and profound implications for the origin, evolution, and future of life on Earth. In coming decades, scientists studying epigenetics may understand how myriad environmental forces affect our health and longevity in specific ways, find cures for many human diseases and conditions, and explain intricate variations in human personality.

The 17th-century philosopher John Locke was convinced that the human mind is a blank slate upon which the environment inscribes personality. With equal self-assurance, others have been convinced that genes orchestrate our development, illnesses, and lifestyles. Yet social scientists failed for decades to explain the mechanisms governing behavioral variations between twins and among family members and culture groups. And biological scientists failed to pinpoint the genetic foundations of many mental illnesses and complex diseases. The central mechanism to explain these complex issues has been found.

I am hardly the first to hail this new field of biology as revolutionary—the fundamental process by which nature and nurture interact. But to me, as an anthropologist long trying to take a middle road in a scientific discipline intractably immersed in nature-versus-nurture warfare, epigenetics is the missing link.

FLOCKING BEHAVIOR IN BIRDS

JOHN NAUGHTON

Newspaper columnist; vice-president, Wolfson College, Cambridge; author, From Gutenberg to Zuckerberg: What You Really Need to Know About the Internet

My favorite explanation is Craig Reynolds's suggestion (first published in 1987) that flocking behavior in birds can be explained by assuming that each bird follows three simple rules: separation (don't crowd your neighbors), alignment (steer toward the average heading of your neighbors), and cohesion (steer toward the average position of your neighbors).[*] That such complex behavior can be accounted for in such a breathtakingly simple way is, well, just beautiful.

---

[*]   C. W. Reynolds, "Flocks, Herds, and Schools: A Distributed Behavioral Model," *Computer Graphics* 21:4 (SIGGRAPH '87 *Conference Proceedings*), 25-34 (1987).

LEMONS ARE FAST

BARRY C. SMITH

*Professor & director, Institute of Philosophy, School
of Advanced Study, University of London; author,*
Questions of Taste: The Philosophy of Wine

When asked to put lemons on a scale between fast and slow, almost everyone says "fast," and we have no idea why. Maybe human brains are just built to respond that way. Probably. But how does that help? It's an explanation of sorts but it seems to be a stopping point, when we wanted to know more. This leads us to ask what we want from an explanation: one that's right, or one that satisfies us? Things that were once self-evident are now known to be false. A straight line is obviously the shortest distance between two points until we think that space is curved. What satisfies our way of thinking need not reflect reality. Why expect a simple theory of a complex world?

Wittgenstein had interesting things to say about what we want from explanations, and he knew different sorts could serve. Sometimes we just need more information; sometimes we need to examine a mechanism, like a valve or a pulley, to understand how it works; sometimes what we need is a way of seeing something familiar in a new light to see it as it really is. He also knew there were times when explanations won't do: "For someone worried by love, an explanatory hypothesis will not help much."*

So, what of the near-universal response to the seemingly meaningless question of whether lemons are fast or slow? To be told that

---

* *Remarks on Frazer's Golden Bough*, trans. Rush Rhees (Oxford, UK: Blackwell, 1967).

our brains are simply built to respond that way doesn't satisfy us. But it's precisely when an explanation leaves us short that it spurs us to greater effort; it's the start of the story, not the end. For the obvious next question to ask is why *are* human brains built this way? What purpose could it serve? And here the phenomenon of automatic associations may give us a deep clue about the way the mind works, because it's symptomatic of what we call cross-modal correspondences: non-arbitrary associations between features in one sense modality with features in another.

There are cross-modal correspondences between taste and shape, between sound and vision, and between hearing and smell, many of which are being investigated by experimental psychologist Charles Spence and philosopher Ophelia Deroy. These unexpected connections are reliable and shared—unlike those in cases of synesthesia, which are idiosyncratic, though individually consistent. And the reason we make these connections in the brain is to give us multiple fixes on objects in the environment that we can both hear and see. It also allows us to communicate elusive aspects of our experience.

We often say that tastes are hard to describe, but when we realize that we can change vocabulary and talk about a taste as round or sharp, new possibilities open up. Musical notes are high or low; sour tastes are high, and bitter notes low. Smells can have low notes and high notes. You can feel low, or be incredibly high. This switching of vocabularies allows us to utilize well-understood sensory modalities to map various possibilities of experience.

Advertisers know this intuitively and exploit cross-modal correspondences between abstract shapes and particular products, or between sounds and sights. Angular shapes conjure up carbonated water, not still—whereas an ice cream called "Frisch" would be thought creamier than one called "Frosch." Notice, too, how many successful companies have names starting with the "K" sound, and how few with "S". These associations set up expectations in the mind that not only help us perceive but may shape our experiences.

And it's not just the vocabularies we use. In his 19th-century tract on the psychology of architecture, Heinrich Wolfflin tells us that it's because we have bodies and are subject to gravity, bending, and balance that we can appreciate the shape of buildings and columns by feeling an empathy for their weight and strain. Physical forms possess a character only because we possess a body.

This idea has led to recent insights into aesthetic appreciation in the work of Chris McManus at University College London. Like all good explanations, it spawns more explanations and further insights. It's another example of how we use the interaction of sensory information to understand and respond to the world around us. So the fact that we all think lemons are fast may be a big part of the reason we're so smart.

FALLING INTO PLACE: ENTROPY AND THE DESPERATE INGENUITY OF LIFE

JOHN TOOBY

Founder of evolutionary psychology; codirector, University of California–Santa Barbara Center for Evolutionary Psychology

The hardest choice I had to make in my early scientific life was whether to give up the beautiful puzzles of quantum mechanics, nonlocality, and cosmology for something equally arresting: to work instead on reverse-engineering the code that natural selection built into the programs that make up our species' circuit architecture. In 1970, the surrounding cultural frenzy and geopolitics made first steps toward a nonideological and computational understanding of our evolved design, "human nature," seem urgent. The recent rise of computer science and cybernetics made it seem possible. The almost complete avoidance of, and hostility to, evolutionary biology by behavioral and social scientists that had nearly neutered those fields made it seem necessary.

What finally pulled me over was that the theory of natural selection was itself such an extraordinarily beautiful and elegant inference engine. Wearing its theoretical lenses was a permanent revelation, populating the mind with chains of deductions that raced like crystal lattices through supersaturated solutions. Even better, it starts from first principles (such as set theory and physics), so much of it is non-optional.

But still, from the vantage point of physics, beneath natural selection there remained a deep problem in search of an explanation: The world given to us by physics is unrelievedly bleak. It blasts us when it is not burning us or invisibly grinding our cells and macromolecules until we are dead. It wipes out planets, habitats, labors, those we love, ourselves. Gamma-ray bursts wipe out entire galactic regions; supernovae, asteroid impacts, super-volcanos, and ice ages devastate ecosystems and end species. Epidemics, strokes, blunt-force trauma, oxidative damage, protein cross-linking, thermal-noise-scrambled DNA—all are random movements away from the narrowly organized set of states that we value, into increasing disorder. The second law of thermodynamics is the recognition that physical systems tend to move toward more probable states, and in so doing tend to move away from less probable states (organization) on their blind toboggan ride toward maximum disorder.

Entropy, then, poses the problem: How are living things at all compatible with a physical world governed by entropy, and, given entropy, how can natural selection lead, over the long run, to the increasing accumulation of functional organization in living things? Living things stand out as an extraordinary departure from the physically normal (for example, the Earth's metal core, lunar craters, or the solar wind). What sets all organisms—from blackthorn and alder to egrets and otters—apart from everything else in the universe is that woven through their designs are staggeringly unlikely arrays of finely tuned interrelationships—a high order that is highly functional. Yet as highly ordered physical systems, organisms should tend to slide rapidly back toward a state of maximum disorder or maximum probability. As the physicist Erwin Schrödinger put it, "It is by avoiding the rapid decay into the inert state of 'equilibrium' that an organism appears so enigmatic."*

The quick answer, normally palmed off on creationists, is true as far as it goes, but far from complete: The Earth is not a closed

---

* *What is Life?* Introduction (1944).

system; organisms are not closed systems, so entropy still increases globally (consistent with the second law of thermodynamics) while (sometimes) decreasing locally in organisms. This permits, but does not explain, the high levels of organization found in life. Natural selection, however, can (correctly) be invoked to explain order in organisms, including the entropy-delaying adaptations that keep us from oxidizing immediately into a puff of ash.

Natural selection is the only known counterweight to the tendency of physical systems to lose rather than grow functional organization—the only natural physical process that pushes populations of organisms uphill (sometimes) into higher degrees of functional order. But how could this work, exactly?

It is here that, along with entropy and natural selection, the third of our trio of truly elegant scientific ideas can be adapted to the problem: Galileo's brilliant concept of frames of reference, which he used to clarify the physics of motion.

The concept of entropy was originally developed for the study of heat and energy, and if the only kind of real entropy was the thermodynamic entropy of energy dispersal, then we (life) wouldn't be possible. But with Galileo's contribution, one can consider multiple kinds of order (improbable physical arrangements), each being defined with respect to a distinct frame of reference.

There can be as many kinds of entropy as there are meaningful frames of reference. Organisms are defined as self-replicating physical systems. This creates a frame of reference that defines its kind of order in terms of causal interrelationships that promote the replication of the system (replicative rather than thermodynamic order). Indeed, organisms must be physically designed to capture undispersed energy, and like hydroelectric dams using waterfalls to drive turbines, they use this thermodynamic entropic flow to fuel their replication, spreading multiple copies of themselves across the landscape.

Entropy sometimes introduces copying errors into replication, but injected disorder in replicative systems is self-correcting. By definition, the less well-organized are worse at replicating themselves and so are removed from the population. In contrast, copying errors that increase functional order (replicative ability) become more common. This inevitable ratchet effect in replicators is natural selection.

Organisms exploit the trick of deploying different entropic frames of reference in many diverse and subtle ways, but the underlying point is that what is naturally increasing disorder (moving toward maximally probable states) for one frame of reference inside one physical domain can be harnessed to decrease disorder with respect to another frame of reference. Natural selection picks out and links different entropic domains (e.g., cells, organs, membranes) that each impose their own proprietary entropic frames of reference locally. When the right ones are associated with each other, they do replicative work through harnessing various types of increasing entropy to decrease other kinds of entropy in ways useful for the organism. For example, oxygen diffusion from the lungs to the bloodstream to the cells is the entropy of chemical mixing—falling toward more probable high-entropy states but increasing order from the perspective of replication promotion.

Entropy makes things fall, but life ingeniously rigs the game so that when they do, they often fall into place.

WHY THINGS HAPPEN

PETER ATKINS

Emeritus professor of chemistry, University of Oxford;
author, Reactions: The Private Life of Atoms

There is a wonderful simplicity in the view that events occur because things get worse. I have in mind the second law of thermodynamics and the fact that all natural change is accompanied by an increase in entropy. Although that is in my mind, I understand those words in terms of the tendency of matter and energy to disperse in disorder. Molecules of a gas undergo ceaseless virtually random motion and spread into the available volume. The chaotic thermal motion of atoms in a block of hot metal jostles their neighbors into motion, and as the energy spreads into the surroundings, so the block cools. All natural change is at root a manifestation of this simple process—that of dispersal in disorder.

The astonishing feature of this perception of natural change is that the dispersal can generate order: Through dispersal in disorder, structure can emerge. All it needs is a device that can link in to the dispersal, and just as a plunging stream of water can be harnessed and used to drive construction, so the stream of dispersal can be harnessed. Overall, there is an increase in disorder as the world progresses, but locally structures, including cathedrals and brains, dinosaurs and dogs, piety and evil deeds, poetry and diatribes, can be thrown up as local abatements of chaos.

Take, for instance, an internal combustion engine. The spark results in the combustion of the hydrocarbon fuel, with the generation of smaller water and carbon dioxide molecules that tend to disperse and in so doing drive down a piston. At the same time, the energy released in the combustion spreads into the surroundings.

The mechanical design of the engine harnesses these dispersals, and, through a string of gears, that harnessing can be used to build from bricks a cathedral. Thus dispersal results in a local structure, even though, overall, the world has sunk a little more into disorder.

The fuel might be our dinner, which, as it is metabolized, releases molecules and energy, which spread. The analog of the gears in a vehicle is the network of biochemical reactions within us, and instead of a pile of bricks molded into a cathedral, amino acids are joined together to generate the intricate structure of a protein. Thus, as we eat, so we grow. We, too, are local abatements of chaos driven into being by the generation of disorder elsewhere.

Is it then too fanciful to imagine intellectual creativity, or just plain inconsequential reverie, as being driven likewise? At some kind of notional rest, the brain is a hive of electric and synaptic activity. The metabolic processes driven by the digestion of food can result in the ordering not of brick into cathedral, not of amino acid into protein, but current into concept, artistic work, foolhardy decision, scientific understanding.

Even that other great principle, natural selection, can be regarded as an extraordinarily complex reticulated unwinding of the world, with the changes that take place in the biosphere and its evolution driven ultimately by descent into disorder. Is it then any wonder that I regard the second law as a great enlightenment? That from so simple a principle great consequences flow is, in my view, a criterion of the greatness of a scientific principle. No principle, I think, can be simpler than that things get worse, and no consequences greater than the universe of all activity, so surely this law is the greatest of all.

WHY WE FEEL PRESSED FOR TIME

ELIZABETH DUNN

Social psychologist, University of British Columbia

Recently I found myself on the side of the road picking gravel out of my knee and wondering how I'd ended up there. I had been biking from work to meet a friend at the gym, pedaling frantically to make up for being a few minutes behind schedule. I knew I was going too fast, and when I hit a patch of loose gravel while careening through a turn, my bike slid out from under me. How had I gotten myself in this position? Why was I in such a rush?

I thought I knew the answer. The pace of life is increasing; people are working more and relaxing less than they did fifty years ago—at least that's the impression one gets from the popular media. But as a social psychologist, I wanted to see the data. As it turns out, there is very little evidence that people nowadays are working more and relaxing less than they did in earlier decades. In fact, some of the best studies suggest just the opposite. So why do people report feeling so pressed for time?

A beautiful explanation for this puzzling phenomenon was recently offered by Sanford DeVoe of the University of Toronto and Jeffrey Pfeffer of Stanford. They argue that as time becomes worth more and more money, time is seen as scarcer. Scarcity and value are perceived as conjoined twins; when a resource—from diamonds to drinking water—is scarce, it is more valuable, and vice versa. So when our time becomes more valuable, we feel as though we had less of it. Surveys around the world have shown that people with higher incomes report feeling more pressed for time—though there are other plausible reasons for this, including the fact that the affluent often work longer hours, leaving them with less free time.

DeVoe and Pfeffer proposed, however, that simply *perceiving* oneself as affluent might be sufficient to generate feelings of time pressure. Going beyond past correlational analyses, they used controlled experiments to put this causal explanation to the test.[*] In one experiment, DeVoe and Pfeffer asked 128 undergraduates to report the total amount of money they had in the bank. All the students answered the question using an 11-point scale, but for half the students, the scale was divided into $50 increments, ranging from $0–$50 to over $500, whereas for the others the scale was divided into much larger increments, ranging from $0–$500 to over $400,000. Most undergraduates using the $50-increment scale circled a number near the top, leaving them with the sense that they were relatively well-off. And this seemingly trivial manipulation led them to feel that they were rushed, pressed for time, and stressed out. Just *feeling* affluent led students to experience the same sense of time pressure reported by genuinely affluent individuals. Using other methods, researchers have confirmed that increasing the perceived economic value of time increases its perceived scarcity.

If feelings of time-scarcity stem in part from the sense that time is highly valuable, then one of the best things we can do to reduce this sense of pressure may be to give our time away. Indeed, new research suggests that giving time away to help others can actually alleviate feelings of time pressure. Companies like Home Depot provide their employees with opportunities to volunteer their time to help others, potentially reducing feelings of time stress and burnout. And Google encourages employees to use 20 percent of their time on their own pet projects, whether or not these have payoff potential. Although some of them resulted in economically

* S. E. DeVoe & J. Pfeffer, "Time Is Tight: How Higher Economic Value of Time Increases Feelings of Time Pressure," *Jour. Appl. Psychol.* 96, 665–76 (2011).

valuable products, like Gmail, the greatest value of this program might lie in reducing employees' sense that their time is scarce.

DeVoe and Pfeffer's work can help account for important cultural trends. Over the past fifty years, feelings of time pressure have risen dramatically in North America, despite the fact that weekly hours of work have stayed fairly level and weekly hours of leisure have climbed. This apparent paradox may be explained in no small part by the fact that incomes have increased substantially during the same period. This causal effect may also help to explain why people walk faster in wealthy cities like Tokyo and Toronto than they do in cities like Nairobi and Jakarta. And at the level of the individual, this explanation suggests that as incomes grow over the course of one's life, time seems increasingly scarce. Which means that as my career develops, I might have to force myself to take those turns a little slower.

WHY THE SUN STILL SHINES

BART KOSKO

Professor of electrical engineering and law, University of Southern California; author, Noise

One of the deepest explanations has to be why the sun still shines— and thus why the sun has not long since burned out, as do the fires of everyday life. That had to worry some of the sun gazers of old as they watched campfires and forest fires burn through their life cycles. It worried the 19th-century scientists who knew that gravity alone could not account for the likely long life of the sun.

It sure worried me when I first thought about it as a child.

The explanation of hydrogen atoms fusing into helium was little comfort. It came at the height of the duck-and-cover cold-war paranoia in the early 1960s after my father had built part of the basement of our new house into a nuclear bomb shelter. The one-room shelter came complete with reinforced concrete and metal windows and a deep freezer packed with homemade TV dinners. The sun burned so long and so brightly because there were in effect so many mushroom-cloud-producing thermonuclear hydrogen-bomb explosions going off inside it and because there was so much hydrogen-bomb-making material in the sun. The explosions were just like the hydrogen bomb explosions that could scorch the Earth and even incinerate the little bomb shelter if they went off close enough.

The logic of the explanation went well beyond explaining the strategic equilibrium of a nuclear Mexican standoff on a global scale. The good news that the sun would not burn out anytime soon came with the bad news that the sun would certainly burn out in a few billion years. But first it would engulf the molten Earth in its red-giant phase.

The same explanation said further that in cosmic due course all the stars would burn out or blow up. There is no free lunch in the heat and light that results when simpler atoms fuse into slightly more complex atoms and when mass transforms into energy. There would not even be stars for long. The universe would go dark and get ever closer to absolute-zero cold. The result would be a faint white noise of sparse energy and matter. Even the black holes would, over eons, burn out or leak out into the near nothingness of an almost perfect faint white noise. That steady-state white noise will have effectively zero information content. It will be the last few steps in a staggeringly long sequence of irreversible nonlinear steps or processes that make up the evolution of the universe. So there will be no way to figure out the lives and worlds that preceded it, even if something arose that could figure.

The explanation of why the sun still shines is as deep as it gets. It explains doomsday.

BOSCOVICH'S EXPLANATION OF ATOMIC FORCES

CHARLES SIMONYI
*Creator, WYSIWYG word processor; cofounder, Intentional
Software; former director of application development and
chief software architect, Microsoft Corporation*

An example of how amazing insight can spring from simple considerations is the explanation of atomic forces by the 18th-century Jesuit polymath Roger Boscovich.

One of the great philosophical arguments at the time took place between the adherents of Descartes who—following Aristotle—thought that forces can only be the result of immediate contact, and those who followed Newton and believed in his concept of force acting at a distance. Newton was the revolutionary here, but his opponents argued—with some justification—that "action at a distance" brought back into physics "occult" explanations that do not follow from the clear understanding that Descartes demanded. Boscovich, a forceful advocate of the Newtonian point of view, turned the question around: Let's understand exactly what happens during the interaction that we would call "immediate contact."

His arguments are easy to understand and extremely convincing. Let's imagine two bodies, one traveling at a speed of, say, 6 units, the other at a speed of 12, with the faster body catching up with the slower along the same straight path. When the two bodies collide, by conservation of the quantity of motion, both should continue after collision along the same path, each with a speed of 9 units in the case of inelastic collision (or, in case of elastic collision, for a brief period right after the collision).

But how did the velocity of the faster body come to be reduced from 12 to 9 and that of the slower body increased from 6 to 9? Clearly, the time interval for the change in velocities cannot be zero, for then, argued Boscovich, the instantaneous change in speed would violate the law of continuity. Furthermore, we would have to say that at the moment of impact, the speed of one body is simultaneously 12 and 9, which is patently absurd.

It is therefore necessary for the change in speed to take place in a small yet finite amount of time. But with this assumption, we arrive at another contradiction. Suppose, for example, that after a small interval of time the speed of the faster body is 11 and that of the slower body is 7. But this would mean that they aren't moving at the same velocity, and the front surface of the faster body would advance through the rear surface of the slower body, which is impossible because we have assumed that the bodies are impenetrable. It therefore becomes apparent that the interaction must take place immediately before the impact of the two bodies and that this interaction can only be a repulsive one, because it is expressed in the slowing down of one body and the speeding up of the other.

Moreover, this argument is valid for arbitrary speeds, so one can no longer speak of definite dimensions for the particles (namely, the atoms) that were until now thought of as impenetrable. An atom should be viewed, rather, as a point source of force, with the force emanating from it acting in some complicated fashion that depends on distance.

According to Boscovich, when bodies are far apart, they act on each other through a force corresponding to the gravitational force, which is inversely proportional to the square of the distance. But with decreasing distance, this law must be modified, because, in accordance with the above considerations, the force changes sign and must become a repulsive force. Boscovich even plotted fanciful traces of how the force should vary with distance in which the force changed sign several times, hinting at the existence of minima in the potential and the existence of stable bonds between the particles, or atoms.

With this idea, Boscovich not only offered a new picture for interactions in place of the Aristotelian-Cartesian theory based on immediate contact, but also presaged our understanding of the structure of matter, especially that of solid bodies.

BIRDS ARE THE DIRECT DESCENDANTS OF DINOSAURS

GREGORY S. PAUL

Independent researcher; author, The Princeton
Field Guide to Dinosaurs

The most graceful example of an elegant scientific idea in one of my fields of expertise is the idea that dinosaurs were tachyenergetic—that they were endotherms with the high internal-energy production and high aerobic-exercise capacity typical of birds and mammals that can sustain long periods of intense activity. Although not dependent on it, the high-powered-dinosaur idea meshes with the hypothesis that birds are the direct descendants of dinosaurs—that birds are flying dinosaurs, much as bats are flying mammals.

The sense that the tachyenergetic idea makes cannot be over-emphasized, nor can the degree to which it revolutionized a big chunk of our understanding of evolution and 230 million years of Earth history, relative to what was thought from the mid-1800s to the 1960s. Until then, it was generally presumed that dinosaurs were a dead-end collection of bradyenergetic reptiles that could achieve high levels of activity for only brief bursts; even walking at a rate of 5 mph requires high respiratory capacity, beyond that of reptiles, who must plod along at a mile per hour or so if they are moving over a long distance. Birds were seen as a distinct and feathery group in which energy inefficiency evolved in order to power flight. Although the latter hypothesis was not inherently illogical, it was divergent from the evolution of bats, whose high aerobic capacity was already present in their furry ancestors.

I first learned of "warm-blooded" dinosaurs in my senior year of high school, via a blurb in *Smithsonian Magazine* about Robert Bakker's article in *Nature* in the summer of 1972. As soon as I read it, it just clicked. I had been illustrating dinosaurs in accord with the reptilian consensus, but it was a bad fit, because dinosaurs are so obviously constructed like birds and mammals, not crocs and lizards. About the same time, John Ostrom, who also had a hand in discovering dinosaur endothermy, was presenting the evidence that birds are aerial versions of avepod dinosaurs—a concept so obvious that it should have become the dominant thesis back in the 1800s.

For a quarter century, the hypotheses were highly controversial—the one regarding dinosaur metabolics especially so—and some of the first justifications were flawed. But the evidence has piled up. Growth rings in dinosaur bones show that they grew at a fast pace not achievable by reptiles. Their tracks showed that they walked at steady speeds too high for bradyaerobes. Many small dinosaurs were feathery. And polar dinosaurs, birds, and mammals were living through blizzardy Mesozoic winters that excluded ectotherms.

Because of the dinorevolution, our understanding of the evolution of the animals that dominated the continents is far closer to the truth than it was. Energy-efficient amphibians and reptiles dominated the continents for only 70 million years in the later portion of the Paleozoic, the era that had begun with trilobites and nothing on land. For the last 270 million years, higher-powered albeit less energy-efficient tachyenergy has reigned supreme on land, starting with protomammalian therapsids near the end of the Paleozoic. When therapsids went belly-up early in the Mesozoic (the survivors of the group being the then-all-small mammals), they were not replaced by lower-powered dinosaurs for the next 150 million years but by dinosaurs that quickly took aerobic-exercise capacity to even greater levels.

The unusual avian respiratory complex is so effective that some birds fly as high as airliners, but the system did not evolve for flight. That's because the skeletal apparatus for operating air-sac-ventilated lungs first developed in flightless avepod dinosaurs for terrestrial purposes (some researchers, but by no means all, offer low global oxygen levels as the selective factor). So the basics of avian energetics appeared in predacious dinosaurs and only later were used to achieve powered flight. Rather like how internal combustion engines happened to make powered human flight practical, rather than having been developed to do so.

COMPLEXITY OUT OF SIMPLICITY

BRUCE HOOD

Director of the Bristol Cognitive Development Centre,
University of Bristol, UK; author, The Self Illusion:
How the Social Brain Creates Identity

As a scientist dealing with complex behavioral and cognitive processes, my deep and elegant explanation comes not from psychology (which is rarely elegant) but from the mathematics of physics. For my money, Fourier's theorem has all the simplicity and yet more power than other familiar explanations in science. Stated simply, any complex pattern, whether in time or space, can be described as a series of overlapping sine waves of multiple frequencies and various amplitudes.

I first encountered Fourier's theorem when I was a PhD student in Cambridge working on visual development. There I met Fergus Campbell, who in the 1960s had demonstrated that not only was Fourier's theorem an elegant way of analyzing complex visual patterns, but it was also biologically plausible. This insight was later to become a cornerstone of various computational models of vision. But why restrict the analysis to vision?

In effect, any complex physical event can be reduced to the mathematical simplicity of sine waves. It doesn't matter whether it is Van Gogh's "Starry Night," Mozart's Requiem, Chanel's No. 5, Rodin's "Thinker," or a Waldorf salad. Any complex pattern in the environment can be translated into neural patterns that, in turn, can be decomposed into the multitude of sine-wave activity arising from the output of populations of neurons.

Maybe I have some physics envy, but to quote Lord Kelvin,

"Fourier's theorem . . . is not only one of the most beautiful results of modern analysis but may be said to furnish an indispensable instrument in the treatment of nearly every recondite question in modern physics."* You don't get much higher praise than that.

---

* *Treatise on Natural Philosophy* (Cambridge: Cambridge University Press, 1879), p. 54.

RUSSELL'S THEORY OF DESCRIPTIONS

A. C. GRAYLING

*Philosopher; master, New College of the Humanities,
London; supernumerary fellow, St. Anne's College, Oxford;
author,* The Good Book: A Humanist Bible

My favorite example of an elegant and inspirational theory in philosophy is Bertrand Russell's theory of descriptions. It did not prove definitive, but it prompted richly insightful trains of inquiry into the structure of language and thought.

In essence, Russell's theory turns on the idea that there is logical structure beneath the surface forms of language, which analysis brings to light; and when this structure is revealed we see what we are actually saying, what beliefs we are committing ourselves to, and what conditions have to be satisfied for the truth or falsity of what is thus said and believed.

One example Russell used to illustrate the idea is the assertion that "the present king of France is bald," said when there is no king of France. Is this assertion true or false? One response might be to say that it is neither, since there is no king of France at present. But Russell wished to find an explanation for the falsity of the assertion that did not dispense with bivalence in logic—that is, the exclusive alternative of truth and falsity as the only two truth-values.

He postulated that the underlying form of the assertion consists in the conjunction of three logically more basic assertions: (a) there is something that has the property of being king of France, (b) there is only one such thing (this takes care of the implication of the definite article "the"), and (c) that thing has the further property of being bald. In the symbolism of first-order predicate calculus, which Russell took to be the properly unambiguous rendering of the assertion's logical form (I omit strictly correct bracketing, so as not to clutter):

$$(Ex)Kx \ \& \ [(y)Ky \rightarrow y = x] \ \& \ Bx$$

which is pronounced "There is an x such that x is K; and for anything y, if y is K then y and x are identical"—this deals logically with "the," which implies uniqueness—"and x is B," where K stands for "has the property of being king of France" and B stands for "has the property of being bald." "E" is the existential quantifier "there is . . ." or "there is at least one . . ." and "(y)" stands for the universal quantifier "for all" or "any."

One can now see that there are two ways in which the assertion can be false; one is if there is no x such that x is K, and the other is if there is an x but x is not bald. By preserving bivalence and stripping the assertion to its logical bones Russell has provided what Frank Ramsey wonderfully called "a paradigm of philosophy."

To the irredeemable skeptic about philosophy, all this doubtless looks like "drowning in two inches of water," as the Lebanese say; but in fact it is in itself an exemplary instance of philosophical analysis, and it has been very fruitful as the ancestor of work in a wide range of fields, from the contributions of Wittgenstein and W. V. Quine to research in philosophy of language, linguistics, psychology, cognitive science, computing, and artificial intelligence.

FEYNMAN'S LIFEGUARD

TIMO HANNAY

Managing director, Digital Science, Macmillan Publishers Ltd.; former publisher, Nature.com; co-organizer, SciFoo

I would like to propose not only a particular explanation but also a particular exposition and exponent: Richard Feynman's lectures on quantum electrodynamics (QED) delivered at the University of Auckland in 1979. These are surely among the very best ever delivered in the history of science.

For a start, the theory is genuinely profound, having to do with the behavior and interactions of those (apparently) most fundamental of particles, photons and electrons. And yet it explains a huge range of phenomena, from the reflection, refraction, and diffraction of light to the structure and behavior of electrons in atoms and their resultant chemistry. Feynman may have been exaggerating when he claimed that QED explains all of the phenomena in the world "except for radioactivity and gravity," but only slightly.

Let me give a brief example. Everyone knows that light travels in straight lines—except when it doesn't, such as when it hits glass or water at anything other than a right angle. Why? Feynman explains that light always takes the path of least time from point to point, and he uses the analogy of a lifeguard racing along a beach to save a drowning swimmer. (This being Feynman, the latter is, of course, a beautiful girl.) The lifeguard could run straight to the water's edge and then swim diagonally along the coast and out to sea, but this would result in a long time spent swimming, which is slower than running on the beach. Alternatively, he could run to the water's edge at the point nearest to the swimmer, and dive in there. But this makes the total distance covered longer than it needs to be. The optimum, if his aim is to reach the girl as quickly as possible, is somewhere in between these two extremes. Light, too, takes such a path of least time from point to point, which is why it bends when passing between different materials.

He goes on to reveal that this is actually an incomplete view. Using the so-called *path integral formulation* (although he avoids that ugly term), Feynman explains that light actually takes every conceivable path from one point to another but most of these cancel each other out, and the net result is that it appears to follow only the single path of least time. This also happens to explain why uninterrupted light (along with everything else) travels in straight lines—so fundamental a phenomenon that surely very few people even consider it to be in need of an explanation. While at first sight such a theory may seem preposterously profligate, it achieves the welcome result of minimizing that most scientifically unsatisfactory of all attributes, arbitrariness.

My amateurish attempts at compressing and conveying this explanation have perhaps made it sound arcane. But on the contrary, a second reason to marvel is that it is almost unbelievably simple and intuitive. Even I, an innumerate former biologist, came away not merely with a vague appreciation that some experts somewhere had found something novel but with the conviction that I was able to share directly in this new conception of reality. Such an experience is all too rare in science generally, but in the abstract, abstruse world of quantum physics it is all but unknown. The main reason for this perspicacity was the adoption of a visual grammar (those famous Feynman diagrams) and an almost complete eschewal of hardcore mathematics (the fact that the spinning vectors central to the theory actually represent complex numbers seems almost incidental). Though the world it introduces is as unfamiliar as can be, it makes complete sense on its own bizarre terms.

THE LIMITS OF INTUITION

BRIAN ENO

Artist; composer; musician; recording producer, U2,
Coldplay, Talking Heads, Paul Simon

We sometimes tend to think that ideas and feelings arising from our intuition are intrinsically superior to those achieved by reason and logic. Intuition—the "gut"—becomes deified as the Noble Savage of the mind, fearlessly cutting through the pedantry of reason. Artists, working from intuition much of the time, are especially prone to this belief. A couple of experiences have made me skeptical.

The first is a question that Wittgenstein used to pose to his students. It goes like this: You have a ribbon, which you want to tie around the center of the Earth (let's assume Earth to be a perfect sphere). Unfortunately, you've tied the ribbon a bit too loose; it's a meter too long. The question is this: If you could distribute the resulting slack—the extra meter—evenly around the planet so the ribbon hovered just above the surface, how far above the surface would it be?

Most people's intuitions lead them to an answer in the region of a minute fraction of a millimeter. The actual answer is almost 16 centimeters. In my experience, only two sorts of people *intuitively* get close to this: mathematicians and dressmakers. I still find it rather astonishing. In fact, when I heard it as an art student, I spent most of one evening calculating and recalculating it, because my intuition was screaming incredulity.

Not many years later, at the Exploratorium in San Francisco, I had another shock-to-the-intuition. I saw for the first time a computer demonstration of John Conway's *Life*. For those of you who don't know it, it's a simple grid with dots that are acted on according to an equally simple and totally deterministic set of rules. The rules decide which dots will live, die, or be born in the next step. There are no tricks, no creative stuff, just the rules. The whole system is so transparent that there should be no surprises at all, but in fact there are plenty: The complexity and "organic-ness" of the evolution of the dot patterns completely beggars prediction. You change the position of one dot at the start and the whole story turns out wildly differently. You tweak one of the rules a tiny bit and there's an explosion of growth or instant Armageddon. You just have no (intuitive) way of guessing which it's going to be.

These two examples elegantly demonstrate the following to me: (a) "Deterministic" doesn't mean "predictable"; (b) we aren't good at intuiting the interaction of simple rules with initial conditions (and the bigger point here is that the human brain may be intrinsically limited in its ability to intuit certain things—like quantum physics and probability, for example); and (c) intuition is not a quasi-mystical voice from outside ourselves speaking through us but a sort of quick-and-dirty processing of our prior experience (which is why dressmakers get it when the rest of us don't). That processing tool sometimes produces impressive results at astonishing speed, but it's worth reminding ourselves now and again that it can also be totally wrong.

THE HIGGS MECHANISM

LISA RANDALL
Physicist, Harvard University; author, Knocking on
Heaven's Door: How Physics and Scientific Thinking
Illuminate the Universe and the Modern World

The beauty of science—in the long run—is its lack of subjectivity. So answering the question "What is your favorite deep, beautiful, or elegant explanation" can be disturbing to a scientist, since the only objective words in the question are "what," "is," "or," and (in an ideal scientific world) "explanation." Beauty and elegance do play a part in science but are not the arbiters of truth. But I will admit that simplicity, which is often confused with elegance, can be a useful guide to maximizing explanatory power.

As for the question, I'll stick to an explanation that I think is extremely nice and relatively simple (though subtle), and which might even be verified within the year. That is the Higgs mechanism, named after the physicist Peter Higgs, who developed it. The Higgs mechanism is probably responsible for the masses of elementary particles, such as the electron. If the electron had zero mass (like the photon), it wouldn't be bound into atoms and none of the structure of our universe would be present.

In any case, experiments have measured the masses of elementary particles and they don't vanish. We know they exist. The problem is that these masses violate the underlying symmetry structure we know to be present in the physical description of these particles. More concretely, if elementary particles had mass from the get-go, the theory would make ridiculous predictions about very energetic particles; for example, it would predict interaction probabilities greater than one.

So here is a significant puzzle. How can particles have masses that have physical consequences and can be measured at low energies but act as if they don't have masses at high energies, when predictions would become nonsensical? That is what the Higgs mechanism tells us. We don't yet know for certain that it is indeed responsible for the origin of elementary particle masses, but no one has found an alternative satisfactory explanation.

One way to understand the Higgs mechanism is in terms of what is known as spontaneous symmetry-breaking, which I'd say is itself a beautiful idea. A spontaneously broken symmetry is broken by the actual state of nature, but not by the physical laws. For example, if you sit at a dinner table and use the glass on your right, so will everyone else. The dinner table is symmetrical—you have a glass on your right and also your left. Yet everyone chooses the glass on the right and thereby spontaneously breaks the left-right symmetry that would otherwise be present.

Nature does something similar. The physical laws describing an object called a Higgs field respect the symmetry of nature. Yet the actual state of the Higgs field breaks the symmetry. At low energy, it takes a particular value. This non-vanishing Higgs field is somewhat akin to a charge spread throughout the vacuum (the state of the universe with no actual particles). Particles acquire their masses by interacting with these "charges." Because this value appears only at low energies, particles effectively have masses only at these energies, and the apparent bottleneck to elementary particle masses is apparently resolved.

Keep in mind that the Standard Model of Particle Physics has worked extremely well, even though we do not yet know whether the Higgs mechanism is correct. We don't need to know about the Higgs mechanism to know that particles have masses and to make many successful predictions with the Standard Model. But the Higgs mechanism is essential to explaining how those masses can arise in a sensible theory. The Standard Model's success none-theless illustrates another beautiful idea essential to all of physics, which is the concept of an "effective theory." The idea is simply that you can focus on measurable quantities when making predictions and leave understanding the source of those quantities to later research when you have better precision.

Fortunately that time has now come for the Higgs mechanism, or at least the simplest implementation, which involves a particle called the Higgs boson. The Large Hadron Collider at CERN, near Geneva, should have a definitive result within the year on whether this particle exists. **[Author's note:** Just before this book went to press, the discovery was announced. Now that a Higgs boson has been found, its properties can be measured to verify whether or not it conforms to the simplest expectations or a more elaborate implementation of the Higgs mechanism.] If confirmed, it will demonstrate that the Higgs mechanism is correct and will furthermore tell us what is the underlying structure responsible for spontaneous symmetry-breaking and spreading "charge" throughout the vacuum. The Higgs boson would furthermore be a new type of particle (a fundamental boson, for those versed in physics terminology) and would be in some sense a new type of force. Admittedly, this is all pretty subtle and esoteric. Yet I (and much of the theoretical physics community) find it beautiful, deep, and elegant.

Symmetry is great. But so is symmetry-breaking. Over the years, many aspects of particle physics were first considered ugly and then considered elegant. Subjectivity in science goes beyond communities to individual scientists. And even those scientists change their minds over time. That's why experiments are critical. As difficult as they are, results are much easier to pin down than the nature of beauty.

THE MIND THINKS IN EMBODIED METAPHORS

SIMONE SCHNALL

Director, Cambridge Embodied Cognition and Emotion Laboratory; University Lecturer, Department of Social and Developmental Psychology, Cambridge, UK

Philosophers and psychologists grappled with a fundamental question for quite some time: How does the brain derive meaning? If thoughts consist of the manipulation of abstract symbols, just as computers process 0s and 1s, then how are such abstract symbols translated into meaningful cognitive representations? This so-called symbol-grounding problem has now been largely overcome, because many findings from cognitive science suggest that the brain does not translate incoming information into abstract symbols in the first place. Instead, sensory and perceptual inputs from everyday experience are taken in their modality-specific form, and they provide the building blocks of thoughts.

British empiricists such as Locke and Berkeley long ago recognized that cognition is inherently perceptual. But following the cognitive revolution in the 1950s, psychology treated the computer as the most appropriate model to study the mind. Now we know that a brain does not work like a computer. Its job is not to store or process information; instead, its job is to drive and control the actions of the brain's large appendage, the body. A new revolution is taking shape, considered by some to bring an end to cognitivism, and ushering in a transformed kind of cognitive science—namely, an *embodied* cognitive science.

The basic claim is that the mind thinks in embodied metaphors.

Early proponents of this idea were linguists, such as George Lakoff, and in recent years social psychologists have been conducting the relevant experiments, providing compelling evidence. But it does not stop here; there is also a reverse pathway. Because thinking is for doing, many bodily processes feed back into the mind to drive action.

Consider the following recent findings that relate to the basic spatial concept of verticality. Because moving around in space is a common physical experience, concepts such as "up" or "down" are immediately meaningful relative to one's own body. The concrete experience of verticality serves as a perfect scaffold for comprehending abstract concepts, such as morality: Virtue is up, whereas depravity is down. Good people are "high-minded" and "upstanding" citizens, whereas bad people are "underhanded" and the "low life" of society. Recent research by Brian Meier, Martin Sellbom, and Dustin Wygant illustrated that research participants are faster to categorize moral words when presented in an up location and immoral words when presented in a down location. Thus people intuitively relate the moral domain to verticality; however, Meier and colleagues also found that people who do not recognize moral norms—namely, psychopaths—fail to show this effect.*

People not only think of all things good and moral as up, but they also think of God as up and the Devil as down. Further, those in power are conceptualized as being high up relative to those over whom they hover and exert control, as shown by Thomas Schubert.† All the empirical evidence suggests that there is indeed a conceptual dimension that leads up, both literally and metaphorically. This vertical dimension that pulls the mind up to consider-

---

* "Failing to take the moral high ground: Psychopathy and the vertical representation of morality," *Personality & Individual Differences* 43, 757–67 (2007).

† "Your highness: Vertical positions as perceptual symbols of power," *J. Personality & Soc. Psychol.* 89, 1–21 (2005).

ing what higher power there might be is deeply rooted in the basic physical experience of verticality.

Verticality not only influences people's representation of what is good, moral, and divine, but movement through space along the vertical dimension can even change their moral actions. Lawrence Sanna, Edward Chang, Paul Miceli, and Kristjen Lundberg recently demonstrated that manipulating people's location along the vertical dimension can turn them into more "high-minded" and "upstanding" citizens. They found that people in a shopping mall who had just moved up an escalator were more likely to contribute to a charity donation box than people who had moved down on the escalator. Similarly, research participants who had watched a film depicting a view from high above—namely, flying over clouds seen from an airplane window—subsequently showed more cooperative behavior than participants who had watched a more ordinary, and less "elevating," view from a car window. Thus being physically elevated induced people to act on "higher" moral values.*

The growing recognition that embodied metaphors provide a common language of the mind has led to fundamentally different ways of studying how people think. For example, under the assumption that the mind functions like a computer, psychologists hoped to figure out how people think by observing how they play chess or memorize lists of random words. From an embodied perspective, it is evident that such scientific attempts were doomed to fail. It is increasingly clear that cognitive operations of any creature, including humans, have to solve certain adaptive challenges of the physical environment. In the process, embodied metaphors are the building blocks of perception, cognition, and action. It doesn't get much more simple and elegant than that.

---

* "Rising Up to Higher Virtues: Experiencing Elevated Physical Height Uplifts Prosocial Actions," *J. Exp. Soc. Psychol.*, 47:2, 472-6 (2011).

METAPHORS ARE IN THE MIND

BENJAMIN K. BERGEN

Associate professor, cognitive science, University of California–San Diego

I study language, and in my field there have been a couple of game-changing explanations over the centuries. One of them explains how languages change over time. Another explains why all languages share certain characteristics. But my favorite is the one that originally got me hooked on language and the mind: It's an explanation of metaphor.

When you look closely at how we use language, you find that a lot of what we say is metaphorical—we talk about certain things as though they were other things. We describe political campaigns as horse races: "Senator Jones has *pulled ahead*." Morality is cleanliness: "That was a *dirty* trick." And understanding is seeing: "New finding *illuminates* the structure of the universe."

People have known about metaphor for a long time. Until the end of the 20th century, almost everyone agreed on one particular explanation, neatly articulated by Aristotle. Metaphor was seen as a strictly linguistic device—a kind of catchy turn of phrase—in which you call one thing by the name of another thing it's similar to. This is probably the definition of metaphor you learned in high school English. According to this view, you can metaphorically say that "Juliet is the sun" if, and only if, Juliet and the sun are similar—for instance, if they are both particularly luminous.

But in their 1980 book *Metaphors We Live By*, George Lakoff and Mark Johnson proposed an explanation for metaphorical language that flouted this received wisdom. They reasoned that if metaphor is just a free-floating linguistic device based on similarity, then you should be able to metaphorically describe anything in terms of anything else it's similar to. But Lakoff and Johnson observed that real metaphorical language, as actually used, isn't haphazard at all. Instead, it's systematic and coherent.

It's systematic in that you don't just metaphorically describe anything as anything else. Instead, it's mostly abstract things that you describe in terms of concrete things. Morality is more abstract than cleanliness. Understanding is more abstract than seeing. And you can't reverse the metaphors. While you can say "He's clean" to mean he has no criminal record, you can't say "He's moral" to mean that he bathed recently. Metaphor is unidirectional.

Metaphorical expressions are also coherent with one another. Take the example of understanding and seeing. There are lots of relevant metaphorical expressions: for example, "I see what you mean" and "Let's shed some light on the issue" and "Put his idea under a microscope and see if it actually makes sense." And so on. While these are totally different metaphorical expressions—they use completely different words—they all coherently cast certain aspects of understanding in terms of specific aspects of seeing. You always describe the understander as the seer, the understood idea as the seen object, the act of understanding as seeing, the understandability of the idea as the visibility of the object, and so on. In other words, the aspects of seeing that you use to talk about aspects of understanding stand in a fixed mapping to one another.

These observations led Lakoff and Johnson to propose that there was something going on with metaphor that was deeper than just the words. They argued that the metaphorical expressions in language are really only surface phenomena, organized and generated by mappings in people's minds. For them, the reason metaphorical language exists and is systematic and coherent is that people think metaphorically. You don't just talk about understanding as seeing; you think about understanding as seeing. You don't just talk about morality as cleanliness; you think about morality as cleanliness.

And it's because you think metaphorically—because you systematically map certain concepts onto others in your mind—that you speak metaphorically. The metaphorical expressions are merely (so to speak) the tip of the iceberg.

As explanations go, this one covers all the bases. It's elegant in that it explains messy and complicated phenomena in terms of something much simpler—a structured mapping between two conceptual domains in the mind. It's powerful in that it explains things other than metaphorical language: Recent work in cognitive psychology shows that people think metaphorically, even in the absence of metaphorical language (affection as warmth, morality as cleanliness). The conceptual-metaphor explanation suggests that we understand abstract concepts like affection or morality by metaphorically mapping them onto more concrete concepts. In terms of utility, the conceptual-metaphor explanation has generated extensive research in a variety of fields; linguists have documented the richness of metaphorical language and explored its diversity across the globe, psychologists have tested its predictions in human behavior, and neuroscientists have searched the brain for its physical underpinnings. And finally, the conceptual metaphor explanation is transformative—it does away with the accepted idea that metaphor is just a linguistic device based on similarity. In an instant, it made us rethink more than 2,000 years of received wisdom. This isn't to say that the conceptual-metaphor explanation doesn't have its weaknesses or that it's the final word in the study of metaphor. But it's an explanation that casts a huge shadow. So to speak.

THE PIGEONHOLE PRINCIPLE

JON KLEINBERG

*Tisch University Professor of computer science, Cornell University;
coauthor (with David Easley)*, Networks, Crowds, and
Markets: Reasoning About a Highly Connected World

Certain facts in mathematics feel as though they contain a kind of compressed power—they look innocuous and mild-mannered when you first meet them, but they're dazzling when you see them in action. One of the most compelling examples is the pigeonhole principle.

Here's what the pigeonhole principle says. Suppose a flock of pigeons lands in a group of trees and there are more pigeons than trees. Then after all the pigeons have landed, at least one of the trees contains more than one pigeon.

This fact sounds obvious, and it is: There are simply too many pigeons, so they can't each get their own tree. Indeed, if this were the end of the story, it wouldn't be clear why this is a fact that deserves to be named or noted. But to appreciate the pigeonhole principle, you have to see some of the things you can do with it.

So let's move on to a fact that doesn't look nearly as straightforward. The statement itself is intriguing, but what's more intriguing is the effortless way it will follow from the pigeonhole principle. Here's the fact: Sometime in the past 4,000 years there have been two people in your family tree—call them A and B—with the property that A was an ancestor of B's mother and also an ancestor of B's father. Your family tree has a loop, where two branches growing upward from B come back together at A—in other words, there's a set of parents in your ancestry who are blood relatives of each other, thanks to this relatively recent shared ancestor A.

It's worth mentioning a couple of things here. First, the "you" in the previous paragraph is genuinely you, the reader. Indeed, one of the interesting features of this fact is that I can make such assertions about you and your ancestors despite not even knowing who you are. Second, the statement doesn't rely on any assumptions about the evolution of the human race or the geographic sweep of human history. Here, in particular, are the only assumptions I'll need:

1. Everyone has two biological parents.

2. No one has children after the age of a hundred.

3. The human race is at least 4,000 years old.

4. At most, a trillion human beings have lived in the past 4,000 years. (Scientists' actual best estimate for (4) is that roughly 100 billion human beings have ever lived in all of human history; I'm bumping this up to a trillion just to be safe.)

All four assumptions are designed to be as uncontroversial as possible; and even then, a few exceptions to the first two assumptions and an even larger estimate in the fourth would only necessitate some minor tweaking to the argument.

Now, back to you and your ancestors. Let's start by building your family tree going back 40 generations: you, your parents, their parents, and so on, 40 steps back. Since each generation lasts, at most, 100 years, the last 40 generations of your family tree all take place within the past 4,000 years. (In fact, they almost surely take place within just the past 1,000 or 1,200 years, but remember that we're trying to be uncontroversial.)

We can view a drawing of your family tree as a kind of "org chart," listing a set of jobs or roles that need to be filled by people. That is, someone needs to be your mother, someone needs to be your father, someone needs to be your mother's father, and so forth, going back up the tree. We'll call each of these an "ancestor role"—it's a job that exists in your ancestry, and we can talk about this job regardless of who actually filled it. The first generation back in your family tree contains two ancestor roles, for your two parents. The second contains four ancestor roles, for your grandparents; the third contains eight roles, for your great-grandparents. Each level you go back doubles the number of ancestor roles that need to be filled, so if you work out the arithmetic, you'll find that forty generations in the past you have more than a trillion ancestor roles that need to be filled.

At this point, it's time for the pigeonhole principle to make its appearance. The most recent 40 generations of your family tree all took place within the past 4,000 years, and we decided that, at most, a trillion people lived during this time. So there are more ancestor roles (over a trillion) than there are people to fill these roles (at most a trillion). This brings us to the crucial point: At least two roles in your ancestry must have been filled by the same person. Let's call this person A.

Now that we've identified A, we're basically done. Starting from two different roles that A filled in your ancestry, let's walk back down the family tree toward you. These two walks downward from A have to first meet each other at some ancestor role lower down in the tree, filled by a person B. Since the two walks are meeting for the first time at B, one walk arrived via B's mother and the other arrived via B's father. In other words, A is an ancestor of B's mother and also an ancestor of B's father, just as we wanted to conclude.

Once you step back and absorb how the argument works, you can appreciate a few things. First, in a way it's more a fact about simple mathematical structures than it is about people. We're taking a giant family tree—yours—and trying to stuff it into the past 4,000 years of human history. It's too big to fit, so certain people have to occupy more than one position in it.

Second, the argument has what mathematicians like to call a nonconstructive aspect. It never really gave you a recipe for finding A and B in your family tree; it convinced you that they must be there, but very little more.

And finally, I like to think of it as a typical episode in the lives of the pigeonhole principle and all the other quietly powerful statements that dot the mathematical landscape—a band of understated little facts that seem frequently to show up at just the right time and, without any visible effort, clean up an otherwise messy situation.

WHY PROGRAMS HAVE BUGS

MARTI HEARST

Computer scientist, University of California–Berkeley, School of Information; author, Search User Interfaces

From the earliest days of computer programming up through the present, we have been faced with the unfortunate reality that the field does not know how to design error-free programs.

Why can't we tame the writing of computer programs to emulate the successes of other areas of engineering? Perhaps the most lyrical thinker to address this question is Frederick Brooks, author of *The Mythical Man-Month.* (If one notes that this unfortunately titled book was first published in 1975, it's easier to ignore the sexist language littering it; the points Brooks made thirty-seven years ago are almost all accurate today, except the assumption that all programmers are "he.")

Espousing the joys of programming, Brooks writes:

> The programmer, like the poet, works only slightly removed from pure thought-stuff. He builds his castles in the air, from air, creating by exertion of the imagination. Few media of creation are so flexible, so easy to polish and rework, so readily capable of realizing grand conceptual structures. . . . Yet the program construct, unlike the poet's words, is real in the sense that it moves and works, producing visible outputs separate from the construct itself. It prints results, draws pictures, produces sounds, moves arms. The magic of myth and legend has come true in our time.

But this magic comes with the bite of its flip side:

In many creative activities the medium of execution is intractable. Lumber splits; paints smear; electrical circuits ring. These physical limitations of the medium constrain the ideas that may be expressed, and they also create unexpected difficulties in the implementation.

. . . Computer programming, however, creates with an exceedingly tractable medium. The programmer builds from pure thought-stuff: concepts and very flexible representations thereof. Because the medium is tractable, we expect few difficulties in implementation; hence our pervasive optimism. Because our ideas are faulty, we have bugs; hence our optimism is unjustified.

Just as there is an arbitrarily large number of ways to arrange the words in an essay, a staggering variety of different programs can be written to perform the same function. The universe of possibility is too wide open, too unconstrained, to permit elimination of errors.

There are additional compelling causes of programming errors, most important the complexiting of autonomously interacting independent systems with unpredictable inputs, often driven by even more unpredictable human actions interconnected on a worldwide network. But in my view, the beautiful explanation is the one about unfettered thought-stuff.

CAGEPATTERNS

HANS-ULRICH OBRIST

Curator, Serpentine Gallery, London; author, Ai Weiwei
Speaks; *coauthor (with Rem Koolhaas),* Project Japan:
Metabolism Talks; *editor,* A Brief History of Curating

In art, the title of a work can often be its first explanation. And in this context I am thinking especially of the titles of Gerhard Richter. In 2006, when I visited Richter in his studio in Cologne, he had just finished a group of six corresponding abstract paintings to which he gave the title *Cage*.

There are many correlations between Richter's paintings and the compositions of John Cage. In a book about the *Cage* series, Robert Storr has traced them from Richter's attendance of a Cage performance at the Festum Fluxorum Fluxus in Düsseldorf in 1963 to analogies in their artistic processes. Cage has often applied chance procedures in his compositions, notably with the use of the I Ching. Richter, in his abstract paintings, also intentionally allows effects of chance. In these paintings, he applies the oil paint on the canvas with a large squeegee. He selects the colors on the squeegee, but the actual trace the paint leaves on the canvas is to a large extent the outcome of chance. The result then forms the basis for Richter's decisions on how to continue with the next layer. In such an inclusion of "controlled chance," an artistic similarity between Cage and Richter can be found. In addition to the reference to John Cage, Richter's title *Cage* also has a visual association, as the six paintings have a hermetic, almost impermeable appearance. The title points to different layers of meaning.

Beyond Richter's abstract paintings, analogies to Cage can be found in other of his works. His book *Patterns* is my favorite book of 2011. It shows Richter's experiment of taking an image of his *Abstract Painting [CR: 724-4]* and dividing it vertically into strips: first 2, then 4, 8, 16, 32, 64, 128, 256, 512, 1,024, 2,048, up to 4,096 strips. This methodology leads to 8,190 strips. Throughout the process, the strips become thinner and thinner. The experiment then leads to the strips being mirrored and repeated, which leads to a diversity of patterns. The outcomes are 221 patterns published on 246 double-page images. In *Patterns*, Richter set the precise rules, but he didn't manipulate the outcome, so the pictures are again an interaction of a defined system and chance.

Patterns is one of many outstanding art books Richter has done over the last couple of years, such as *Wald* (2008), or *Ice* (1981), which includes a special layout of the artist with his stunning photos of a trip to the Antarctic. The layout of those books is composed of intervals with different arrangements of the photos but also blank spaces—like pauses. Richter told me that his layout has to do with music, Cage, and silence.

In 2007, Richter designed a 20-meter-high arched stained-glass window to fill the south transept of Cologne Cathedral. The *Cologne Cathedral Window* comprises 11,000 hand-blown squares of glass in 72 colors derived from the palette of the original Medieval glazing that was destroyed during the Second World War. Half of the squares were allotted by a random generator, the other half were like a mirror image to them. Control is once more ceded here to some extent, suggesting his interest in Cage's ideas to do with chance and the submission of the individual will to forces beyond one's control. "Coincidences are only useful," Richter has told me, "because they've been worked out—that means either eliminated or allowed or emphasized."

In Halberstadt, a performance of Cage's piece "ORGAN[2]/ASLSP" (1987) recently took place. "ASLSP" stands for "as slow as possible." Cage has not further specified this instruction, so that each performance of the score will be different. The actual performance will take 639 years to be completed. The slowness in Cage's piece is an essential aspect for our time. With globalization and the Internet, all processes have been accelerated to a speed in which no time for critical reflection remains. The present "Slow movement" thus advises us to take time for well-chosen decisions together with a more locally oriented approach. The idea of slowness is one of the many aspects that continue to make Cage most relevant for the 21st century.

Richter's concise title, *Cage*, can be unfolded into an extensive interpretation of these abstract paintings (and of other works)—but, one can say, the short form already contains everything. The title, like an explanation of a phenomenon, unlocks the works, describing their relation to one of the most important cultural figures of the 20th century, John Cage, who shares with Richter the great themes of chance and uncertainty.

THE TRUE ROTATIONAL SYMMETRY OF SPACE

SETH LLOYD

Professor of quantum mechanical engineering, MIT;
author, Programming the Universe

The following deep, elegant, and beautiful explanation of the true rotational symmetry of space comes from the late Sidney Coleman, as presented to his graduate physics class at Harvard. This explanation takes the form of a physical act that you will perform yourself. Although elegant, the explanation is verbally awkward to explain and physically awkward to perform. It may need to be practiced a few times. So limber up and get ready: You are about to experience in a deep and personal way the true rotational symmetry of space!

At bottom, the laws of physics are based on symmetries, and the rotational symmetry of space is one of the most profound of these symmetries. The most rotationally symmetric object is a sphere. So take a sphere such as a soccer ball or basketball that has on it a mark, logo, or unique lettering at some spot. Rotate the sphere about any axis: The rotational symmetry of space implies that the shape of the sphere is invariant under rotation. In addition, if there is a mark on the sphere, then when you rotate the sphere by 360°, the mark returns to its initial position. Go ahead; try it. Hold the ball in both hands and rotate it by 360° until the mark returns.

That's not so awkward, you may say. But that's because you have not yet demonstrated the true rotational symmetry of space. To demonstrate this symmetry requires fancier moves. Now hold the ball cupped in one hand, palm facing up. Your goal is to rotate the sphere while keeping your palm up. This is trickier, but if Michael Jordan can do it, so can you.

The steps are as follows:

Keeping your palm up, rotate the ball inward toward your body. At 90°, one quarter of a full rotation, the ball is comfortably tucked under your arm.

Keep on rotating in the same direction, palm up. At 180°, half a rotation, your arm sticks out in back of your body to keep the ball cupped in your palm.

As you keep rotating to 270°, three-quarters of a rotation, in order to maintain your palm facing up, your arm sticks awkwardly out to the side, ball precariously perched on top.

At this point, you may feel that it is impossible to rotate the last 90° to complete one full rotation. If you try, however, you will find that you can continue rotating the ball keeping your palm up by raising your upper arm and bending your elbow so that your forearm sticks straight forward. The ball has now rotated by 360°—one full rotation. If you've done everything right, however, your arm should be crooked in a maximally painful and awkward position.

To relieve the pain, continue rotating by an additional 90° to one-and-a-quarter turns, palm up all the time. The ball should now be hovering over your head, and the painful tension in your shoulder should be somewhat lessened.

Finally, like a waiter presenting a tray containing the *pièce de resistance*, continue the motion for the final three-quarters of a turn, ending with the ball and your arm (what a relief!) back in its original position.

If you have managed to perform these steps correctly and without personal damage, you will find that the trajectory of the ball has traced out a kind of twisty figure eight, or infinity sign (∞), in space and has rotated around not once but twice. The true symmetry of space is not rotation by 360° but by 720°.

Although this exercise might seem no more than a fancy and painful basketball move, the fact that the true symmetry of space is rotation not once but twice has profound consequences for the nature of the physical world at its most microscopic level. It implies that "balls," such as electrons, attached to a distant point by flexible and deformable "strings," such as magnetic field lines, must be rotated around twice to return to their original configuration. Digging deeper, the twofold rotational nature of spherical symmetry implies that two electrons, both spinning in the same direction, cannot be placed in the same place at the same time. This exclusion principle in turn underlies the stability of matter. If the true symmetry of space were rotating around only once, then all the atoms of your body would collapse into nothingness in a tiny fraction of a second. Fortunately, however, the true symmetry of space consists of rotating around twice, and your atoms are stable, a fact that should console you as you ice your shoulder.

THE PIGEONHOLE PRINCIPLE REVISITED

CHARLES SEIFE

Professor of journalism, New York University; former writer, Science; *author,* Proofiness: The Dark Arts of Mathematical Deception

Sometimes even the simple act of counting can tell you something profound. One day, back in the late 1990s, when I was a correspondent for *New Scientist* magazine, I got an e-mail from a flack waxing rhapsodic about an extraordinary piece of software. It was a revolutionary data-compression program so efficient it would squash every digital file by 95 percent or more without losing a single bit of data. Wouldn't my magazine jump at the chance to tell the world about the computer program that would make hard drives hold twenty times more information than before?

No, my magazine wouldn't.

No such compression algorithm could possibly exist; it was the algorithmic equivalent of a perpetual-motion machine. The software was a fraud. The reason: the pigeonhole principle.

The pigeonhole principle is a simple counting argument. It says that if you've got *n* pigeons and manage to stuff them into fewer than *n* boxes, then at least one box must have more than one pigeon in it. As blindingly obvious as this is, it's a powerful tool. For example, imagine that the compression software really worked as advertised and every file shrank by a factor of 20 with no loss of fidelity. Every single file 2,000 bits long would be squashed down to a mere 100 bits, and then, when the algorithm was reversed, it would expand back into its original form, unscathed.

When compressing files, you bump up against the pigeonhole

principle. There are many more 2,000-bit pigeons (2^{2000}, to be exact) than 100-bit boxes (2^{100}). If an algorithm stuffs the former into the latter, at least one box must contain multiple pigeons. Take that box—that 100-bit file—and reverse the algorithm, expanding the file into its original 2,000-bit form. You can't! Since there are multiple 2,000-bit files that all wind up being squashed into the same 100-bit file, the algorithm has no way of knowing which one was the true original—it can't reverse the compression.

The pigeonhole principle puts an absolute limit on what a compression algorithm can do. It can compress some files, often dramatically, but it can't compress them all—at least, not if you insist on perfect fidelity.

Counting arguments similar to this one have opened up entire new realms for us to explore. The German mathematician Georg Cantor used a kind of reverse pigeonhole-principle technique to show that it was impossible to fit the real numbers into boxes labeled by the integers—even though there is an infinite number of integers. The almost unthinkable consequence was that there are different levels of infinity. The infinity of the integers is dwarfed by the infinity of the reals, which, in turn, is dwarfed by yet another infinity and another infinity on top of that . . . an infinity of infinities, all unexplored until we learned to count them.

Taking the pigeonhole principle into deep space has an even stranger consequence. A principle in physics, the holographic bound, implies that in any finite volume of space there is only a finite number of possible configurations of matter and energy. If, as cosmologists tend to believe, the universe is infinite, there is an infinite number of visible-universe-size volumes out there—enormous cosmos-size bubbles containing matter and energy. And if space is more or less homogeneous, there's nothing particularly special about the cosmos-size bubble we live in. These assumptions, taken together, lead to a stunning conclusion. Infinite universe-size bubbles, with only a finite number of configurations of the matter and energy in each, mean that there's not just an exact copy of our universe—and our Earth—out there; the transfinite version of the pigeonhole principle states that there's an infinite number of copies of every (technically, "almost every," which has a precise mathematical definition) possible universe. Not only are there infinite copies of you on infinite alternate Earths, there are infinite copies of countless variations on the theme: versions of you with a prehensile tail, versions of you with multiple heads, versions of you that have made a career juggling carnivorous rabbitlike animals in exchange for costume jewelry. Even something as simple as counting one, two, three can lead you to bizarre and unexpected realms.

MOORE'S LAW

RODNEY A. BROOKS

Roboticist; Panasonic Professor of Robotics, emeritus, MIT;
founder, chairman & CTO, Heartland Robotics, Inc.; author,
Flesh and Machines: How Robots Will Change Us

Moore's Law originated in a four-page 1965 magazine article written by Gordon Moore, then at Fairchild Semiconductor and later one of the founders of Intel. In it, he predicted that the number of components on a single integrated circuit would rise from the then-current number of roughly 2^6 to roughly 2^{16} in the following ten years—that is, the number of components would double every year. He based this prediction on four empirical data points and one null data point, fitting a straight line on a graph plotting the log of the number of components on a single chip against a linear scale of calendar years. Intel later amended Moore's Law to say that "the number of transistors on a chip roughly doubles every two years."

Moore's Law is rightly seen as the fundamental driver of the information technology revolution in our world over the last fifty years. Doubling the number of transistors every so often has made our computers twice as powerful for the same price, doubled the amount of data they can store or display, made them twice as fast, made them smaller, made them cheaper, and in general improved them in every possible way by a factor of 2 on a clockwork schedule.

But why does it happen? Automobiles have not obeyed Moore's Law; neither have batteries, nor clothing, nor food production, nor the level of political discourse. All but the last have demonstrably improved due to the influence of Moore's Law, but none has had the same relentless exponential improvements.

The most elegant explanation for what makes Moore's Law possible is that digital logic is all about an abstraction—and, in fact, a one-bit abstraction, a yes/no answer to a question—and that abstraction is independent of physical bulk.

In a world that consists entirely of piles of red sand and piles of green sand, the size of the piles is irrelevant. A pile is either red or green, and you can take away half the pile, and it's still either a pile of red sand or a pile of green sand. And you can take away another half, and another half, and so on, and still the abstraction is maintained. And repeated halving at a constant rate makes an exponential.

That's why Moore's Law works for digital technology and doesn't work for technologies that require physical strength, or physical bulk, or must deliver certain amounts of energy. Digital technology uses physics to maintain an abstraction and nothing more.

Some caveats do apply:

1. In his short paper, Moore expressed some doubt as to whether his prediction would hold for linear, rather than digital, integrated circuits, pointing out that by their nature "such elements require the storage of energy in a volume" and that the volume would necessarily be large.

2. It does matter when you get down to piles of sand with just one grain, and then technology has to shift and you need to use some new physical property to define the abstraction. Such technology shifts have happened again and again in the maintenance of Moore's Law over almost fifty years.

3. The idea does not explain the sociology of how Moore's Law is implemented or what determines the time constant of a doubling, but it does explain why exponentials are possible in this domain.

COSMIC COMPLEXITY

JOHN C. MATHER

*Senior astrophysicist, Observational Cosmology
Laboratory, NASA's Goddard Space Center; coauthor
(with John Boslough),* The Very First Light

What explains the extraordinary complexity of the observed universe, on all scales from quarks to the accelerating universe? My favorite explanation (which I certainly did not invent) is that the fundamental laws of physics produce natural instability, energy flows, and chaos. Some call the result the Life Force, some note that the Earth is a living system itself (Gaia, a "tough bitch," according to the late Lynn Margulis), and some conclude that the observed complexity requires a supernatural explanation (of which we have many). But my dad was a statistician (of dairy cows) and he told me about cells and genes and evolution and chance when I was very small. So a scientist must look for the explanation of how nature's laws and statistics brought us into conscious existence. And how it is that seemingly improbable events are happening all the time.

Well, physicists have countless examples of natural instability, in which energy is released to power change from simplicity to complexity. One of the most common is that cooling water vapor below the freezing point produces snowflakes, no two alike and all complex and beautiful. We see it often, so we're not amazed. But physicists have observed so many kinds of these changes from one structure to another (we call them phase transitions) that the Nobel Prize in 1992 could be awarded for understanding the mathematics of their common features.

Now for a few examples of how the laws of nature produce the instabilities that lead to our own existence. First, the Big Bang (what an insufficient name!) apparently came from an instability, in which the "false vacuum" eventually decayed into the ordinary vacuum we have today, plus the most fundamental particles we know, the quarks and leptons. So the universe as a whole started with an instability. Then a great expansion and cooling happened, and the loose quarks, finding themselves unstable too, bound themselves together into today's less elementary particles—protons and neutrons—liberating a little energy and creating complexity. Then the expanding universe cooled some more, and neutrons and protons, no longer kept apart by immense temperatures, found themselves unstable and formed helium nuclei. Then a little more cooling, and atomic nuclei and electrons were no longer kept apart and the universe became transparent. Then a little more cooling, and the next instability began: Gravitation pulled matter together across cosmic distances to form stars and galaxies. This instability is described as a "negative heat capacity," in which extracting energy from a gravitating system makes it hotter—clearly the second law of thermodynamics does not apply here. (This is the physicists' version of e. e. cummings's notion of "the wonder that's keeping the stars apart.") Then the next instability is that hydrogen and helium nuclei fuse to release energy and make stars burn for billions of years. And then at the end of the fuel source, stars become unstable and explode and liberate their chemical elements into space. And because of that, on planets like Earth, sustained energy flows support the development of additional instabilities and all kinds of complex patterns. Gravitational instability pulls the densest materials into the core of the Earth, leaving a thin skin of water and air, and makes the interior churn incessantly as heat flows outward. And the heat from the sun, received mostly near the equator and flowing toward the poles, supports the complex atmospheric and oceanic circulations.

And because of all that, the physical Earth is full of natural chemical laboratories, concentrating elements here, mixing them there, raising and lowering temperatures, ceaselessly experimenting with uncountable events where new instabilities can arise. At least one of them was the new experiment called Life. Now that we know there are at least as many planets as there are stars, it's hard to imagine that nature's ceaseless experimentation would not be able to produce Life elsewhere—but we don't know for sure.

And Life went on to cause new instabilities, constantly evolving, with living things in an extraordinary range of environments, changing the global environment, with boom-and-bust cycles, with predators for every kind of prey, with criminals for every possible crime, with governments to prevent them, and instabilities of the governments themselves.

One of the instabilities is that humans demand new weapons and new products of all sorts, leading to serious investments in science and technology. So the natural/human world of competition and combat is structured to lead to advanced weaponry and cell phones. So here we are in 2012, with people writing essays and wondering whether their descendants will be artificial lifeforms traveling back into space. And pondering what the origins of those forces of nature are that give rise to everything. The Dutch theoretical physicist Erik Verlinde has argued that gravitation, the one force that has so far resisted our efforts at a quantum description, is not even a fundamental force but is itself a statistical force, like osmosis.

What an amazing turn of events! But after all I've just said, I should not be surprised a bit.

THE GAIA HYPOTHESIS

SCOTT SAMPSON

Dinosaur paleontologist and science communicator; author,
Dinosaur Odyssey: Fossil Threads in the Web of Life

For my money, the deepest, most beautiful scientific explanation is
the Gaia hypothesis, the idea that Earth's physical and biological
processes are inextricably interwoven to form a self-regulating system. This notion—the 1965 brainchild of chemist James Lovelock,
further developed with microbiologist Lynn Margulis—proposes
that air (atmosphere), water (hydrosphere), Earth (geosphere), and
life (biosphere) interact to form a single evolving system capable
of maintaining environmental conditions consistent with life.
Lovelock initially put forth the Gaia hypothesis to explain how
life on Earth has persisted for almost 4 billion years despite a 30
percent increase in the sun's intensity over that interval.

But how does Gaia work? Lovelock and Margulis demonstrated
that, lacking a conscious command-and-control system, Gaia uses
feedback loops to track and adjust key environmental parameters.
Take oxygen, a highly reactive by-product of life, generated and
continually replenished by photosynthetic algae and plants. The
present atmospheric concentration of oxygen is about 21 percent.
A few percentage points lower and air-breathing life-forms could
not survive; a few percentage points higher and terrestrial ecosystems would become overly combustible, prone to conflagration.
According to the Gaia hypothesis, oxygen-producing organisms
have used feedback loops to maintain atmospheric oxygen between
these narrow limits for hundreds of millions of years.

Similar arguments, backed by an ever-growing body of
research, can be made for other atmospheric constituents, as

well as for global surface temperature, ocean salinity, and other key environmental metrics. Although the Gaia hypothesis highlights cooperation at the scale of the biosphere, researchers have documented multiple examples showing how cooperation at one level could evolve through competition and natural selection at lower levels. Initially criticized by serious scientists as New Age mumbo-jumbo, Lovelock's radical notion has increasingly been incorporated into scientific orthodoxy, and key elements are now often taught as "Earth systems science." One timely lesson resulting at least in part from Gaian research is that food-web complexity, including higher species diversity, tends to enhance ecological and climate stability.

So, while Earth may inhabit a Goldilocks zone—neither too close nor too far from the sun—life's rampant success on this "pale blue dot" cannot be ascribed to luck alone. Life has had a direct hand in ensuring its own persistence.

Science has not yet fully embraced the Gaia hypothesis. And it must be admitted that as an explanation, the idea remains incomplete. The insights cascading from Gaia are unquestionably deep and beautiful, uniting the whole of the biosphere and Earth's surface processes into a single, emergent, self-regulating system. But this explanation has yet to achieve the third milestone defined in this year's *Edge* Question: elegance. The Gaia hypothesis lacks the mathematical precision of Einstein's $e = mc^2$. No unified theory of Earth and life has been presented to explain why life stabilizes more than it destabilizes.

Evolutionary biologist W. D. Hamilton once compared Lovelock's insights to those of Copernicus, adding that we still await the Newton who will define the laws of this grand, seemingly improbable relationship. Hamilton himself became deeply engrossed in seeking an answer to this question, developing a computer model that seemed to show how stability and productivity could increase in tandem. Were it not for his untimely death, he might have emerged as that modern-day Newton.

The cultural implications of Gaia also continue to be debated. Arguably the most profound implication of Lovelock's idea is that Earth considered as a whole has many qualities of an organism. But is Gaia actually alive, akin to a single life-form, or is it more accurate to think of her as a planet-size ecosystem? Lynn Margulis argued strongly (and convincingly, to my mind) for the latter view. Margulis, whose work revolutionized evolutionary biology at the smallest and grandest of scales, died last year. Always the hard-nosed scientist, she once said, "Gaia is a tough bitch—a system that has worked for over 3 billion years without people. This planet's surface and its atmosphere and environment will continue to evolve long after people and prejudice are gone."

While not disagreeing with this blunt assessment, I find considerably greater inspiration in Gaian thinking. Indeed, I would go as far as to suggest that this idea can help shift the human perception of nature. In the modernist perspective, the natural world is little more than a collection of virtually infinite resources available for human exploitation. The Gaian lens encourages us to reenvision Earthbound nature as an intertwined, finite whole from which we evolved and in which we remain fully embedded. Here, then, is a deep and beautiful perspective in desperate need of broad dissemination.

THE CONTINUITY EQUATIONS

LAURENCE C. SMITH
Professor of geography, UCLA; author, The World in 2050

These are already familiar to you—at least, in anecdotal form. Almost everyone has heard of the law of conservation of mass (sometimes with the word "matter" in place of "mass") and probably its partner, conservation of energy, too. These laws tell us that for practical, real-world (that is, non-quantum, non–general relativity) phenomena, matter and energy can never be created or destroyed, only shuffled around. That concept has origins at least as far back as the ancient Greeks, was formally articulated in the 18th century (a major advance for modern chemistry), and today underpins virtually every aspect of the physical, life, and natural sciences. Conservation of mass (or matter) is what finally quashed the alchemists' quest to transform lead into gold; conservation of energy is what consigns the awesome power of a wizard's staff to the imaginations of legions of *Lord of the Rings* fans.

The continuity equations take these laws an important step further, by providing explicit mathematical formulations that track the storage and/or transfers of mass (mass continuity) and energy (energy continuity) from one compartment or state to another. As such, they are not really a single pair of equations but are instead written into a variety of forms, ranging from the very simple to the very complex, in order to best represent the physical phenomenon they are supposed to describe. The most elegant forms, adored by mathematicians and physicists, have exquisite detail and are therefore the most complex. A classic example is the set of Navier-Stokes equations—sometimes called the Saint-Venant equations—used to understand the movements and accelerations of fluids. The beauty of Navier-Stokes lies in their explicit partitioning and tracking of mass, energy, and momentum through space and time. However, in practice, such detail also makes these equations difficult to solve, requiring either hefty computing power or simplifying assumptions to be made to the equations themselves.

But the power of the continuity equations is not limited to complex forms comprehensible solely to mathematicians and physicists. A forest manager, for example, might use a simple, so-called mass-balance form of a mass-continuity equation to study her forest by adding up the number, size, and density of trees, determining the rate at which seedlings establish themselves, and then subtracting the trees' mortality rate and number of truckloads of timber removed, so as to learn whether its total wood content (biomass) is increasing, decreasing, or stable. Automotive engineers routinely apply simple energy-balance equations when, for example, designing a hybrid electric car to recapture kinetic energy from its braking system. None of the energy is truly created or destroyed, just recaptured—in this case, from a combustion engine, which got it from breaking apart ancient chemical bonds, which got it from photosynthetic reactions, which got it from the sun. Any remaining energy not recaptured from the brakes is not really lost, of course, but instead transferred to the atmosphere as low-grade heat.

The cardinal assumption behind these laws and equations is that mass and energy are conserved within a closed system. In principle, the hybrid electric car satisfies energy continuity only if its consumption is tracked from start (the sun) to finish (dissipation of heat into the atmosphere). This is a cumbersome calculation, so the process is usually treated as an open system. The metals used in the car's manufacture satisfy mass continuity only if tracked from their source (ores) to landfill. This tracking is more feasible, and such cradle-to-grave resource accounting—a high priority for many environmentalists—is thus more compatible with natural laws than our current economic model, which tends to treat such resource flows as open systems.

Like the car, our planet is, from a practical standpoint, an open system with respect to energy and a closed system with respect to mass. (Although Earth is still being bombarded by meteorites, that input is now small enough to be ignored.) The former is what makes life possible: Without the sun's steady infusion of fresh, external energy, life as we know it would quickly end. An external source is required because although energy cannot be destroyed, it constantly degrades into weaker, less useful forms, in accordance with the second law of thermodynamics. (Consider the hybrid-electric car's brake pads—their dissipated heat is not of much use to anyone.) The openness of this system is two-way, because Earth also streams thermal infrared energy back out to space. Its radiation is invisible to us, but to satellites with "vision" in this range of the electromagnetic spectrum the Earth is a brightly glowing orb, much like the sun.

Interestingly, this closed/open dichotomy is yet another reason that the physics of climate change are unassailable. By burning fossil fuels, we shuffle carbon (mass) out of the subsurface—where it has virtually no interaction with the planet's energy balance—to the atmosphere, where it does. It is well understood that carbon in the atmosphere alters the planet's energy balance; the physics of this has been known since 1893, thanks to the Swedish chemist Svante Arrhenius. Without carbon-based and other greenhouse gases, our planet would be a moribund, ice-covered rock. Greenhouse gases prevent this by selectively altering the Earth's energy balance in the troposphere, the lowest few miles of the atmosphere, where the vast majority of its gases reside—thus raising the amount of thermal infrared radiation that Earth emits.

Because some of this energy streams back down to Earth as well as out to space, the lower troposphere warms to achieve energy balance. Continuity of energy commands this.

Our planet's carbon atoms, however, are stuck here with us forever—continuity of mass commands that, too. The question is, What choices will we make about how extensively and rapidly to shuffle them out of the ground? The physics of natural resources, climate change, and other problems can often be reduced to simple, elegant equations—if only we had tools masterful enough to dictate their solution.

PASCAL'S WAGER

TIM O'REILLY
Founder and CEO of O'Reilly Media

In 1661 or 1662, in his *Pensées*, philosopher and mathematician Blaise Pascal articulated what would come to be known as Pascal's Wager, the question of whether or not to believe in God in the face of the failure of reason and science to provide a definitive answer:

> You must wager. It is not optional. You are embarked. Which will you choose then? . . . You have two things to lose, the true and the good; and two things to stake, your reason and your will, your knowledge and your happiness; and your nature has two things to shun, error and misery. Your reason is no more shocked in choosing one rather than the other, since you must of necessity choose. This is one point settled. But your happiness? Let us weigh the gain and the loss in wagering that God is. Let us estimate these two chances. If you gain, you gain all; if you lose, you lose nothing. Wager, then, without hesitation that He is.

While this proposition of Pascal's is clothed in obscure religious language and on a religious topic, it is a significant and early expression of decision theory. And stripped of its particulars, it provides a simple and effective way to reason about contemporary problems like climate change.

We don't need to be 100 percent sure that the worst fears of climate scientists are correct in order to act. All we need to think about are the consequences of being wrong.

Let's assume for a moment that there is no human-caused climate change, or that the consequences are not dire, and we've made big investments to avert it. What's the worst that happens? In order to deal with climate change:

1. We've made major investments in renewable energy. This is an urgent issue even in the absence of global warming, as the International Energy Agency has now revised the date of "peak oil" to 2020, only eight years from now.

2. We've invested in a potent new source of jobs.

3. We've improved our national security by reducing our dependence on oil from hostile or unstable regions.

4. We've mitigated the enormous off-the-books economic losses from pollution. (China recently estimated these losses as 10 percent of GDP.) We currently subsidize fossil fuels in dozens of ways, by allowing power companies, auto companies, and others to keep environmental costs off the books, by funding the infrastructure for autos at public expense while demanding that railroads build their own infrastructure, and so on.

5. We've renewed our industrial base, investing in new industries rather than propping up old ones. Climate skeptics like Bjorn Lomborg like to cite the cost of dealing with global warming. But these costs are similar to the "costs" incurred by record companies in the switch to digital-music distribution, or the "costs" to newspapers implicit in the rise of the Web. That is, they are costs to existing industries, but they ignore the opportunities for

new industries that exploit the new technology. I have yet to see a convincing case made that the costs of dealing with climate change aren't principally the costs of protecting old industries.

By contrast, let's assume that the climate skeptics are wrong. We face the displacement of millions of people, droughts, floods and other extreme weather, species loss, and economic harm that will make us long for the good old days of the current financial-industry meltdown.

Climate change really is a modern version of Pascal's wager. On one side, the worst outcome is that we've built a more robust economy. On the other, the worst outcome really is Hell. In short, we do better if we believe in climate change and act on that belief, even if we turn out to be wrong.

But I digress. The illustration has become the entire argument. Pascal's wager is not just for mathematicians, nor for the religiously inclined. It is a useful tool for any thinking person.

EVOLUTIONARILY STABLE STRATEGIES

S. ABBAS RAZA

Founding editor, 3quarksdaily.com

My example of a deep, elegant, beautiful explanation in science is John Maynard Smith's concept of an *evolutionarily stable strategy* (ESS). Not only does this wonderfully straightforward idea explain a whole host of biological phenomena, but it also provides a useful heuristic tool to test the plausibility of various types of claims in evolutionary biology—allowing us, for example, to quickly dismiss group-selectionist misconceptions, such as the idea that altruistic acts by individuals can be explained by the benefits that accrue to the species as a whole. Indeed, the idea is so powerful that it explains things I didn't even realize needed explaining until I was given the explanation.

I will now present one such explanation to illustrate the power of ESS. I should note that while Smith developed ESS using the mathematics of game theory (along with collaborators G. R. Price and G. A. Parker), I will attempt to explain the main idea using almost no math.

Think of common animal species like cats, or dogs, or humans, or golden eagles. Why do all of them have (nearly) equal numbers of males and females? Why is there not sometimes 30 percent males and 70 percent females in a species? Or the other way around? Or some other ratio altogether? Why are sex ratios almost exactly fifty-fifty? I, at least, never entertained the question until I read the elegant answer.

Let's consider walruses: They exist in the normal fifty-fifty sex ratio, but most walrus males will die virgins, whereas almost all females will mate. Only a few dominant walrus males monopolize most of the females. So what's the point of having all those extra males around? They take up food and resources, but in the only thing that matters to evolution they are useless, because they do not reproduce. From a species point of view, it would be better and more efficient if only a small proportion of walruses were males and the rest were females; such a species of walrus would make much more efficient use of its resources and, according to the logic of group-selectionists, would soon wipe out the actual existing species of walrus with the inefficient fifty-fifty gender ratio. So why hasn't that happened?

Here's why: because a population of walruses (you can substitute any other species of animals I've mentioned, including humans, for the walruses in this example) with, say, 10 percent males and 90 percent females (or any other non–fifty-fifty ratio) would not be stable over a large number of generations. Why not? In the 10 percent males and 90 percent females of this example, each male is producing about nine times as many children as any female—by successfully mating with, on average, nine females. Imagine such a population. If you were a male in this kind of population, it would be to your evolutionary advantage to produce more sons than daughters, because each son could be expected to produce roughly nine times as many offspring as any of your daughters. Let me run through some numbers to make this clearer: Suppose the average male walrus fathers ninety children, only nine of which will be males and eighty-one females, on average, and the average female walrus bears ten baby walruses, only one of which will be a male and nine of which will be females. OK?

Here's the crux of the matter: Suppose a mutation arose in one of the male walruses—as well it might over a large number of generations—that gave this particular male walrus more Y (male-producing) sperm than X (female-producing) sperm. This gene would spread like wildfire through the described population. Within a few generations, more and more male walruses would have the gene that makes them have more male than female offspring, and soon you would get the fifty-fifty ratio we see in the real world.

The same argument applies to females: Any mutation in a female that caused her to produce more male than female offspring (though sex is determined by the sperm, not the egg, there are other mechanisms the female might employ to affect the sex ratio) would spread quickly in this population, bringing the ratio closer to fifty-fifty with each subsequent generation. In fact, any significant deviation from the fifty-fifty gender ratio will, for this reason, be evolutionarily unstable and through random mutation will soon revert to it. And this is just one example of the deep, elegant, and beautiful explanatory power of ESS.

THE COLLINGRIDGE DILEMMA

EVGENY MOROZOV

Journalist; visiting scholar, Stanford University; Schwartz Fellow, New America Foundation; author, The Net Delusion: The Dark Side of Internet Freedom

In 1980, David Collingridge, an obscure academic at the University of Aston in the UK, published an important book called *The Social Control of Technology*, which set the tone of many subsequent debates about technology assessment. In it, he articulated what has become known as the Collingridge dilemma—the idea that there is always a tradeoff between knowing the impact of a given technology and the ease of influencing its social, political, and innovation trajectories.

Collingridge's basic insight was that we can successfully regulate a given technology when it's still young and unpopular and thus probably still hiding its unanticipated and undesirable consequences—or we can wait and see what those consequences are, but then risk losing control over its regulation. Or as Collingridge himself so eloquently put it: "When change is easy, the need for it cannot be foreseen; when the need for change is apparent, change has become expensive, difficult, and time-consuming." The Collingridge dilemma is one of the most elegant ways to explain many of the complex ethical and technological quandaries—think drones or automated facial recognition—that plague our globalized world.

TRUSTING TRUST

ERNST PÖPPEL

Psychologist; neuroscientist; CEO, Human Science Center, Munich University; author, Mindworks: Time and Conscious Experience

After many years
A little gift to *Edge*
From the first culture.

Using the haiku
Five/seven/five syllables
To express a thought.

Searching for beauty
To explain the unexplained
Why should I do this?

What is my problem?
I don't need explanations!
I'm happy without!

A new morning comes.
I wake up leaving my dreams,
And I don't know why.

I don't understand
Why I can trust my body
In day and in night.

Looking at the moon,
Always showing the same face,
But I don't know why!

Must I explain this?
Some people certainly can.
Beyond my power!

I look at a tree.
But is there in fact a tree?
I trust in my eyes.

But why do I trust?
Not understanding my brain
Being too complex.

Looking for answers,
Searching for explanations,
But living without.

Trust in my percepts
And trust in my memories.
Trust in my feelings.

Where does it come from,
This absolute certainty,
This trust in the world?

Trusting in the future,
Making plans for tomorrow,
Why do I believe?

I have no answer!
Knowledge is not sufficient.
Only questions count.

What is a question?
That is the real challenge!
Finding a new path.

But trust is required.
Believing the new answers,
Hiding in a shadow.

Deep explanations.
Rest in the trust of answers,
Which is unexplained.

Is there a way out?
Evading the paradox?
This answer is no!

The greatest challenge:
Accepting the present,
Giving no answers!

IT JUST IS?

BRUCE PARKER

Visiting professor, Center for Maritime Systems, Stevens Institute of Technology; oceanographer; author, The Power of the Sea: Tsunamis, Storm Surges, Rogue Waves, and Our Quest to Predict Disasters

The concept of an *indivisible component of matter*, something that cannot be divided further, has been around for at least two and a half millennia, first proposed by early Greek and Indian philosophers. Democritus called the smallest indivisible particle of matter *átomos*, meaning "uncuttable." Atoms were also simple, eternal, and unalterable. But in Greek thinking (and generally for about 2,000 years after), atoms lost out to the four basic elements of Empedocles—fire, air, water, earth—which were also simple, eternal, and unalterable but not made up of little particles, Aristotle believing those four elements to be infinitely continuous.

Further progress in our understanding of the world, based on the concept of atoms, had to wait until the 18th century. By that time, the four elements of Aristotle had been replaced by the thirty-three elements of Lavoisier, based on chemical analysis. Dalton then used the concept of atoms to explain why elements always react in ratios of whole numbers, proposing that each element is made up of atoms of a single type and that these atoms can combine to form chemical compounds. Of course, by the early 20th century (through the work of Thomson, Rutherford, Bohr, and many others), it was realized that atoms were not indivisible and thus not the basic units of matter. All atoms were made up of protons, neutrons, and electrons, which took over the title of being the *indivisible components* (basic building blocks) *of matter.*

Perhaps because the Rutherford-Bohr model of the atom is now considered transitional to more elaborate models based on quantum mechanics, or perhaps because it evolved over time from the work of many people (and wasn't a single beautiful proposed law), we have forgotten how much about the world can be explained by the concept of protons, neutrons, and electrons—probably more than any other theory ever proposed. With only three basic particles, one could explain the properties of 118 atoms/elements and the properties of thousands upon thousands of compounds chemically combined from those elements. A rather amazing feat, and certainly making the Rutherford-Bohr model worthy of being considered a "favorite deep, elegant, and beautiful explanation."

Since that great simplification, further developments in our understanding of the physical universe have gotten more complicated, not less. To explain the properties of our three basic particles of matter, we went looking for even-more-basic particles. We ended up needing twelve fermions (six quarks, six leptons) to "explain" the properties of the three previously thought-to-be basic particles (as well as the properties of some other particles that were not known to us until we built high-energy colliders). And we added four other particles, force-carrier particles, to "explain" the four fundamental force fields (electromagnetism, gravitation, strong nuclear interaction, and weak nuclear interaction) that affect those three previously thought-to-be basic particles. Of these sixteen now thought-to-be basic particles, most are not independently observable (at least at low energies).

Even if the present Standard Model of Particle Physics turns out to be true, the question can be asked: "What next?" Every particle, whatever its level in the hierarchy, will have certain properties or characteristics. When we are asked why quarks have a particular electric charge, color charge, spin, or mass, do we simply say, "They just do"? Or do we try to find even-more-basic particles that seem to explain the properties of quarks, leptons, and bosons? And if so, does this continue to still-even-more-basic particles? Could it go on forever? Or at some point, when asked, "Why does this particle have these properties?," would we simply say, "It just does"? At some point, would we have to say that there is no "why" to the universe? It just is.

At what level of our hierarchy of understanding would we resort to saying, "It just is"? The first level (with the least amount of understanding about the world) is religious: the gods of Mount Olympus, each responsible for some worldly phenomenon, or the all-knowing monotheistic god creating the world and making everything work by means truly unknowable to humans. In their theories about how the world worked, Aristotle and other Greek philosophers incorporated the Olympian gods (earth, water, fire, and air were all assigned to particular gods), but Democritus and other philosophers were deterministic and materialistic, and they looked for predictable patterns and simple building blocks that might create the complex world they saw around them. In the evolution of scientific thinking, there have been various "It just is" moments, when an explanation or theory seems to hit a wall, until someone comes along and says, "Maybe not" and goes on to advance our understanding. But as we get to the most basic questions about our universe (and our existence), the "It just is" answer becomes more likely. One basic scientific question is whether truly indivisible particles of nature will ever be found. The accompanying philosophical question is whether there *can be* truly indivisible particles of nature.

At some level, the next group of mathematically derived "particles" may so obviously appear not to be observable/"real" that we will describe them instead as simply entities in a mathematical model that seems to accurately describe the properties of the observable particles in the level above. At which point, the answer to the question of why these particles act as described by this mathematical model would be "They just do." How far down we go with such models will probably depend on how much a new level in the model allows us to explain previously unexplainable observed phenomena or correctly predict new phenomena. (Or perhaps we might be stopped by the model's becoming too complex.)

For determinists still unsettled by the probabilities inherent in quantum mechanics or the philosophical question about what would have come before a Big Bang, it's just one more step toward recognizing the true unsolvable mystery of our universe—recognizing it, but maybe still not accepting it. A new and much better model could still come along.

SUBVERTING BIOLOGY

PATRICK BATESON
Professor of ethology, Cambridge University; coauthor
(with Paul Martin), Design for a Life

Two years ago, I reviewed the evidence on inbreeding in pedigreed dogs. Inbreeding can result in reduced fertility both in litter size and sperm viability, developmental disruption, lower birth rate, higher infant mortality, shorter life span, increased expression of inherited disorders, and reduced immune-system function. The immune system is closely linked to the removal of cancer cells from a healthy body, and, indeed, reduced immune-system function increased the risk of full-blown tumors. These well-documented cases in domestic dogs confirm what is known from many wild populations of other species. It comes as no surprise, therefore, that a variety of mechanisms render inbreeding less likely in the natural world. One such is the choice of unfamiliar individuals as sexual partners.

Despite all the evidence, the story is more complicated than at first appears, and this is where the explanation for what happens has a certain beauty. While inbreeding is generally seen as undesirable, the debate has become much more nuanced in recent years. Purging of genes with seriously damaging effects carries obvious benefits, and this can happen when a population is inbred. Outcrossing, which is usually perceived as advantageous, does carry the possibility that the benefits of purging are undone by introducing new harmful genes into a population. Furthermore, a population adapted to one environment may not do well if crossed with a population adapted to another. So a balance is often struck between inbreeding and outbreeding.

When the life history of a species demands careful nurturing of the offspring, the parents may go to a lot of trouble to mate with the best partner possible. A mate should be not too similar to oneself, but not too dissimilar either. Japanese quail of both sexes prefer first cousins as partners. Subsequent animal studies have suggested that an optimal degree of relatedness is most beneficial to the organism in terms of reproductive success. A study of a human Icelandic population also points to the same conclusion. Couples who are third or fourth cousins have a larger number of grandchildren than more closely or more distantly related partners. Much evidence from humans and nonhuman animals suggests that the choice of a mate depends on experiences in early life, with individuals tending to choose partners who are a bit different but not too different from familiar individuals, who are usually but not always close kin.

The role of early experience in determining sexual and social preferences bears on a well-known finding that humans are extremely loyal to members of their own group. They are even prepared to give up their own lives in defense of those with whom they identify. In sharp contrast, they can behave with lethal aggressiveness toward those who are unfamiliar to them. This suggests, then, a hopeful resolution to the racism and intolerance that bedevils many societies. As people from different countries and ethnic backgrounds become better acquainted with one another, they will be more likely to treat one another well, particularly if the familiarity starts at an early age. If familiarity leads to marriage, the couples may have fewer grandchildren, but that may be a blessing on an overpopulated planet. This optimistic principle, generated by knowledge of how a balance has been struck between inbreeding and outbreeding, subverts biology—but it does hold, for me, considerable beauty.

SEX AT YOUR FINGERTIPS

SIMON BARON-COHEN

*Psychologist; director, Autism Research Centre, Cambridge University;
author,* The Science of Evil: On Empathy and the Origins of Cruelty

We all know males and females are different below the neck. There's growing evidence that there are differences above the neck, too. Looking into the mind reveals that, on average, females develop empathy faster—and, on average, males develop stronger interests in systems, or how things work. These are differences not so much in ability as in cognitive style and patterns of interest. They shouldn't stand in the way of achieving equal opportunities in society or equal representation in all disciplines and fields, but such political aspirations are a separate issue from the scientific observation of cognitive differences.

Looking into the brain also reveals differences. For example, whereas males, on average, have larger brain volume than females, even correcting for height and weight, females on average reach their peak volume of gray and white matter at least a year earlier than males. There's also a difference in the number of neurons in the neocortex: On average, males have 23 million and females 19 million, a 16 percent difference. Looking at other brain regions also shows sex differences: For example, males, on average, have a larger amygdala (an emotion area) and females, on average, a larger planum temporale (a language area). But in all this talk about sex differences, ultimately what we want to know is what gives rise to these differences, and here is where I, at least, enjoy some deep, elegant, and beautiful explanations.

My favorite is fetal testosterone, since extra drops of this special molecule seem to have "masculinizing" effects on the development of the brain and the mind. The credit for this simple idea must go to Charles Phoenix and colleagues at the University of Kansas who proposed it in 1959[*] and to Norman Geschwind and Albert Galaburda at Harvard, who picked it up in the early 1980s. This is not the only masculinizing mechanism (another is the X chromosome), but it is one that has been elegantly dissected.

However, scientists who study the causal properties of fetal testosterone sometimes resort to unethical animal experiments. Take, for example, a part of the amygdala called the medial posterodorsal (MePD) nucleus, which is larger in male rats than in females. If you castrate the poor male rat (thereby depriving him of the main source of his testosterone), the MePD shrinks to the female volume in just four weeks. Or you can do the reverse experiment, giving extra testosterone to a female rat, which makes her MePD grow to the same size as a typical male rat, again in just four weeks.

In humans, we look for more ethical ways of studying how fetal testosterone does its work. You can measure this special hormone in the amniotic fluid that bathes the fetus in the womb. It gets into the amniotic fluid by being excreted by the fetus and so is thought to reflect the levels of this hormone in the baby's body and brain. My Cambridge colleagues and I measured unborn male babies' testosterone in this way and then invited them into an MRI brain scanner some ten years later. In a recent paper in the *Journal of Neuroscience*, our group shows, for example, that the more testosterone there is in the amniotic fluid, the less gray matter in the planum temporale.[†]

---

[*] "Organizing Action of Prenatally Administered Testosterone Propionate on the Tissues Mediating Mating Behavior in the Female Guinea Pig," *Endocrinology* 65:3, 369-82 (1959).

[†] Michael V. Lombardo et al., "Fetal Testosterone Influences Sexually Dimorphic Gray Matter in the Human Brain," *J. Neurosci.* 32: 2, 674-80 (2012).

This fits with an earlier finding we published, that the more testosterone in the amniotic fluid, the smaller the child's vocabulary size at age two.* This helps make sense of a longstanding puzzle about why girls talk earlier than boys and why boys are disproportionately represented in clinics for language delays and disorders, since boys in the womb produce at least twice as much testosterone as girls.

It also helps make sense of the puzzle of individual differences in rate of language development in typical children regardless of their sex: why at two years old some children have huge vocabularies (600 words) and others haven't even started talking. Fetal testosterone is not the only factor involved in language development—so are social influences, since firstborn children develop language faster than later children—but it seems to be a key part of the explanation. And fetal testosterone has been shown to be associated with a host of other sex-linked features, from eye contact to empathy and from detailed attention to autistic traits.

Fetal testosterone is tricky to get your hands on, since the last thing a scientist wants to do is interfere with the delicate homeostasis of the uterine environment. In recent years, a proxy for fetal testosterone has been proposed: the ratio between the second- and fourth-finger lengths, or the 2D:4D ratio. Males have a lower ratio than females in the population, and this is held to be set in the womb and to remain stable throughout one's life. So scientists no longer have to think of imaginative ways to measure the testosterone levels directly in the womb. They can simply take a xerox of someone's hand, palm down, at any time in their life to measure a proxy for levels of testosterone in the womb.

I was skeptical of the 2D:4D measure for a long time, simply

---

*   S. Lutchmaya et al., "Foetal testosterone and vocabulary size in 18- and 24-month-old infants," *Infant Behavior & Development*, 24:4, 418-24 (2002).

because it made little sense that the relative length of your second and fourth fingers should have anything to do with your hormones prenatally. But just last year, in *Proceedings of the National Academy of Sciences*, Zheng and Cohn showed that even in mice paws, the density of receptors for testosterone and estrogen varies in the second and fourth digits, making a beautiful explanation for why your finger-length ratio is directly affected by these hormones.[*] The same hormone that masculinizes your brain is at work at your fingertips.

---

[*] Zhenghui Zheng & Martin J. Cohn, "Developmental basis of sexually dimorphic digit ratios," *PNAS* 108:39, 16289-94 (2011).

WHY DO MOVIES MOVE?

ALVY RAY SMITH
Cofounder, Pixar; digital imagery pioneer

Movies are not smooth. The time between frames is empty. The camera records only twenty-four snapshots of each second of time flow and discards everything that happens between frames—but we perceive it anyway. We see stills, but we perceive motion. How can we explain this? We can ask the same question about digital movies, videos, and videogames—in fact, all modern digital media—so the explanation is rather important, and one of my favorites.

Hoary old "persistence of vision" can't be the explanation. It's real, but it explains only why you don't see the emptiness between frames. If an actor or an animated character moves between frames, then—by persistence of vision—you should see him in both positions: two Humphrey Bogarts, two Buzz Lightyears. In fact, your retinas do see both, one fading out as the other comes in—each frame is projected long enough to ensure this. It's what your brain does with the retinas' information that determines whether you perceive two Bogarts in two different positions or one Bogart moving.

On its own, the brain perceives the motion of an edge, but only if the edge moves not too far, and not too fast, from the first frame to the second. Like persistence of vision, this is a real effect, called *apparent motion*. It's interesting, but it's not the explanation I like so much. Classic cel animation—of the old ink-on-celluloid variety—relies on the apparent-motion phenomenon. The old animators knew intuitively how to keep the successive frames of a movement inside the "not too far, not too fast" boundaries. If they needed to exceed those limits, they had tricks to help us perceive the motion—like actual speed lines and a *poof* of dust to mark the rapid descent of Wile E. Coyote as he steps unexpectedly off a mesa in hot pursuit of that truly wily Road Runner.

Exceed the apparent-motion limits without those animators' tricks and the results are ugly. You may have seen old-school stop-motion animation—such as Ray Harryhausen's classic sword-fighting skeletons in *Jason and the Argonauts*—plagued by an unpleasant jerking motion of the characters. You're seeing double—several edges of a skeleton at the same time—and correctly interpreting it as motion, but painfully so. The edges stutter, or "judder," or "strobe" across the screen—words that reflect the pain inflicted by staccato motion.

Why don't live-action movies judder? (Imagine directing Uma Thurman to stay within not-too-far-not-too-fast limits.) Why don't computer-animated movies à la Pixar judder? And, for contrast, why do videogames, alas, strobe horribly? All are sequences of discrete frames. There's a general explanation that works for all three. It's called *motion blur*, and it's simple and pretty.

Here's what a real movie camera does. The frame it records is not a sample at a single instant, like a Road Runner or a Harryhausen frame. Rather, the camera shutter is open for a short while, called the exposure time. A moving object is moving during that short interval, of course, and is thus smeared slightly across the frame during the exposure time. It's like what happens when you try to take a long-exposure still photo of your child throwing a ball, and his arm is just a blur. But a bug in a still photograph turns out to be a feature for movies. Without the blur, all movies would look as jumpy as Harryhausen's skeletons.

A scientific explanation can become a technological solution. For digital movies—like *Toy Story*—the solution to avoid strobing was derived from the explanation for live-action: Deliberately smear a moving object across a frame along its path of motion. So a character's swinging arm must be blurred along the arc the arm traces as it pivots around its shoulder joint. And the other arm independently must be blurred along *its* arc, often in the opposite direction to the first arm. All that had to be done was to figure out how to do with a computer what a camera does—and, importantly, how to do it efficiently. Live-action movies get motion blur for free, but it costs a lot for digital movies. The solution—by the group now known as Pixar—paved the way for the first digital movie. Motion blur was the crucial breakthrough.

In effect, motion blur shows your brain the path a movement is taking and also its magnitude—a longer blur means a faster motion. Instead of discarding the temporal information about motion between frames, we store it spatially in the frames as a blur. A succession of such frames overlapping a bit—because of persistence of vision—thus paints out a motion in a distinctive enough way that the brain can make the full inference.

Pixar throws thousands of computers at a movie—spending sometimes more than thirty hours on a single frame. On the other hand, a videogame—essentially a digital movie restricted to real time—has to deliver a new frame every thirtieth of a second. It was only seventeen years ago that the inexorable increase in computation speed per unit dollar (described by Moore's Law) made motion-blurred digital movies feasible. Videogames simply haven't arrived yet. They can't compute fast enough to motion-blur. Some give it a feeble try, but the feel of the play lessens so dramatically that gamers turn it off and suffer the judder instead. But Moore's Law still applies, and soon—five years? ten?—even videogames will motion-blur properly and fully enter the modern world.

Best of all, motion blur is just one example of a potent general explanation called the *sampling theorem*. The theorem works when the samples are frames, taken regularly in time to make a movie, or when they're pixels, taken regularly in space to make an image. It works for digital audio, too. In a nutshell, the explanation of smooth motion from unsmooth movies expands to explain the modern media world—why it's even possible. But that would take a longer explanation.

WOULD YOU LIKE BLUE CHEESE WITH IT?

ALBERT-LÁSZLÓ BARABÁSI
Complex network scientist; Distinguished Professor and director of Northeastern University's Center for Complex Network Research; *author,* Bursts: The Hidden Pattern Behind Everything We Do

It would take about 100 years to try the 100,000 recipes carried on Epicurious, the largest recipe portal in the United States. What fascinates me about this number is not how huge it is but how tiny. Indeed, a typical dish has about eight ingredients. Thus the roughly 300 ingredients used in cooking today allow for about a quadrillion distinct dishes. Add to this your choice of deep-freezing, frying, smashing, centrifuging, or blasting your ingredients and you start to see why cooking is a growth industry. It currently uses only a negligible fraction of its resources—less than one in a trillion dishes that culinary combinatorics permits.

Don't you like green eggs and ham? Or why leave this vast *terra incognita* unexplored? Do we simply lack the time to taste our way through this boundless bounty, or is it because most combinations are repugnant? Could there be some rules that explain liking some ingredient combinations and avoiding others? The answer appears to be yes, which leads me to my most flavorful explanation to date.

As we search for evidence to support (or refute) any "laws" that may govern our culinary experiences, we must bear in mind that food sensation is affected by many factors, from color to texture, from temperature to sound. Yet palatability is largely determined by flavor, representing a group of sensations including odors, tastes, freshness, and pungency. This is mainly chemistry, however. Odors are molecules that bind olfactory receptors, tastes are chemicals that stimulate taste buds, freshness or pungency are signaled by chemical irritants in our mouth and throat. Therefore, if we want to understand why we prize some ingredient combinations and loathe others, we have to look at the chemical profile of our recipes.

But how can chemistry tell us which ingredients taste well together? Well, we can formulate two orthogonal hypotheses. First, we may like some ingredients together because their chemistry (henceforth, their flavor) is complementary—what one lacks is provided by the other. The alternative is the polar opposite: Taste is like color matching in fashion—we prefer to pair ingredients that already share some flavor compounds, bringing them in chemical harmony with one another. Before you go on reading, I urge you to stop for a second and ponder which of these you find more plausible.

The first one makes more sense to me: I put salt in my omelet not because the chemical bouquet of the egg shares the salt's only chemical, NaCl, but precisely because it is missing it. Yet lately, chefs and molecular gastronomers are betting on the second hypothesis, and they have even given it a name, calling it the *food-pairing principle*. Its consequences are already on your table. Some contemporary restaurants serve white chocolate with caviar because they share trimethylamine and other flavor compounds, or chocolate and blue cheese because they share at least seventy-three flavor compounds. Yet evidence for food pairing is at best anecdotal, making a scientist like myself ask: Is this more than a myth?

So whom should I trust, my intuition or the molecular gastronomers? And how to really test if two ingredients indeed go well together? Our first instinct was to taste, under controlled conditions, all ingredient pairs. Yet 300 ingredients offered about 44,850 pairs to sample, forcing us to search for smarter ways to settle the question. Having spent the last decade trying to understand the laws governing networks, from the social network to the intricate web of genes governing our cells, my colleagues and I decided to rely on network science. We compiled the flavor components of over 300 ingredients and organized them into a network, linking two ingredients if they shared flavor compounds. We then used the collective intelligence accumulated in the existing body of recipes to test what goes with what. If two common ingredients are almost never combined, like garlic and vanilla, there must be a reason for it; those who tried it may have found it either uninspired or outright repulsive. If, however, two ingredients are combined more often than we would expect, based on their individual popularity, we took that as a sign that they must taste well together. Tomato and garlic are in this category, combined in 12 percent of all recipes.*

The truth is rather Dr. Seussian at the end: We may like some combinations here, but not there. That is, North American and Western European cuisine show a strong tendency to combine ingredients that share chemicals. If you are *here*, serve parmesan with papaya and strawberries with beer. Do not try this *there*, however: East Asian cuisine thrives by avoiding ingredients that share flavor chemicals. So if you hail from Asia, yin/yang is your guiding force: seeking harmony through pairing the polar opposites. Do you like soy sauce with honey? Try them together and you might.

* Yong-Yeol Ahn et al., "Flavor network and the principles of food pairing," *Scientific Reports* 1, article 196, doi:10.1038/srep00196 (2011).

MOTHER NATURE'S LAWS

STUART PIMM

Doris Duke Professor of Conservation Ecology, Nicholas School of the Environment, Duke University; author, A Scientist Audits the Earth

Writing from Sarawak, Alfred Russel Wallace nailed the most important law of living things in a crisp eighteen words: *"Every species has come into existence coincident both in space and time with a pre-existing closely allied species."*

With judicious editing, Wallace could have fit his 1855 "laws of evolution" paper into today's word limits of *PNAS* or *Nature*. We don't find trilobites scattered in the Devonian, Jurassic, and Eocene with nothing in between. The paper screams for an explanation of the bundled generalities of paleontology and biogeography, but the scientific community was asleep at the wheel and barely noticed. A few years later, that absence of notice forced Wallace to send his deep, elegant, and beautiful explanation to Darwin for moral support. Darwin had the same explanation, of course.

What other laws has Mother Nature given us for biological diversity?

The average geographical range of a group of species is very much larger than the median range.
The average of the geographical ranges of 1,684 species of mammals in the New World is 1.8 million km², but half of those species have ranges smaller than 250,000 km²—a seven-to-one ratio. For the region's three main bird groups, the ratio is five and eight times, and for amphibians, forty times. There are many species with small ranges and few with large ranges.

There are more species in the tropics than in temperate regions.
The first explorers to reach the tropics uncovered this law. Rembrandt was painting birds of paradise and marine cone shells in the early 1600s. Wallace went first to the Amazon because collecting novel species was how he earned a living.

Species with small ranges concentrate in places that typically are not where the largest numbers of species live.
This just doesn't make sense. Surely, with more species, one should have more species with large ranges, small ranges, and everything in between. It isn't so. Small-ranged species concentrate in some very special places. About half of all species live in a couple of dozen places that together constitute about 10 percent of the ice-free parts of the planet.

Species with small ranges are rare within those ranges, while those with large ranges are common.
Pardon the language, but Mother Nature is a bitch. You'd think she'd give species with small ranges a break—and make them locally common. Not so. Widespread species tend to be common everywhere, while local ones are rare even where you find them.

What inspired Darwin and Wallace were encounters with places rich in birds and mammals found nowhere else—the Galapagos and the islands of Southeast Asia. There are no such places in Europe. Darwin spent most of *HMS Beagle*'s voyage too far south in South America, while Wallace's first trip was to the Amazon. The Amazon is very rich in species, but it is a striking example of the law that such places rarely have many species with small ranges. (I suspect this cost Wallace dearly, because his sponsors wanted novelty. He found novelty on his next trip, to the East.)

Scientists found widespread species first. Darwin and Wallace were among the first naturalists to encounter the majority of species—those with small geographical ranges concentrated in a few places. Even for well-known groups of species, those with the smallest ranges have been discovered only in the last decades.

What deep, elegant, and beautiful explanation underpins these ineluctably connected laws? There isn't one.

Given the observed distribution of range sizes, the tropics have to have more species simply because they are in the middle of the globe. Sufficiently large ranges must span the middle—that's the only way to fit them in. But they need not be at the ends—temperate or arctic places. Yet middles have more species than ends even when the middles aren't tropical. There are more species in the middle of Madagascar's wet forests, though the northern end (with fewer species) is closer to the equator, for example.

Moreover, warm, wet middles—tropical moist forests—have more species than hotter and drier middles. The correlation of species with warmth and wetness is compelling, but a compelling mechanism is sometimes an illusion.

Small-ranged species can be anywhere—near middles or near ends. They are not. They tend to be on islands (the Galapagos, the Malaysian archipelago), and on "habitat islands"—mountaintops, like the Andes. This fits our ideas on how species form. Alas, they are not on temperate islands and mountains, so Darwin and Wallace had to leave home to be inspired. Except for salamanders: The Appalachians of the Eastern United States seem to have different species under every rock, forming a theoretically obstinate temperate center of endemism unmatched by birds, mammals, plants, or, indeed, other amphibians.

To make matters worse, all this assumes that we know why some species have large ranges and more have small. We do not. In short, we have correlations, special cases, and some special pleading, but elegance is missing. A deep explanation need not be there, of course.

Our ignorance hurts. Concentrations of local, rare species are where human actions drive species to extinction 100 to 1,000 times faster than the natural rate. Yes, we can map birds and mammals and so know where we need to act to save them. But not butterflies, which people love, let alone nematodes. Without explanations, we cannot tell whether the places where we protect birds will also protect butterflies. Unless we understand Mother Nature's laws and extend them to the great majority of species still unknown to science, we may never know what we destroyed.

THE OKLO PYRAMID

KARL SABBAGH
Writer and television producer; author, Remembering
Our Childhood: How Memory Betrays Us

New explanations in science are needed when an observation isn't explicable by current theory. The power of the scientific method lies in the extraordinary richness of understanding that can emerge from an attempt to devise a new explanation. It's like an inverted pyramid, with the first observation—often just a slight departure from the norm—as the point and then ever widening layers of inference, each dependent on a lower layer, until the whole pyramid supplies a satisfying and conclusive explanatory whole.

One of my favorite such explanations dates from 1972, with the observation of a small anomaly in a routine sample of uranium ore from Oklo, a region in the Haut-Ogooué Province of the Central African state of Gabon, which was analyzed in a French laboratory. Rock samples of naturally occurring uranium usually contain two types of uranium atoms, the isotopes U-238 and U-235. Most of the atoms are U-238, but about 0.7 percent are U-235. In fact, to be accurate, the figure is 0.72 percent, but the sample that arrived in France had "only" 0.717 percent, meaning that .003 percent of the expected U-235 atoms were missing.

Such differences in proportion were known to occur only in the artificial surroundings of a nuclear reactor, where U-235 is bombarded with neutrons in a chain reaction that transforms the atoms and leads to the change in the naturally occurring proportions. But this sample had come from a mine in Gabon, and at the time there was no nuclear reactor on the whole continent of Africa, so that couldn't be the explanation. Or could it?

Nearly twenty years earlier, scientists had suggested that somewhere on Earth the conditions might once have existed for a uranium deposit to act like a natural nuclear-fission reactor. They proposed three necessary conditions:

1. The size of the deposit should be greater than the average length that fission-inducing neutrons travel, which is about 70 centimeters.

2. U-235 atoms must be present in a greater abundance than in natural rocks today, as much as 3 percent instead of 0.72 percent.

3. There must be what is called in a nuclear reactor a moderator, a substance that "blankets" the emitted neutrons and slows them down so they're more apt to induce fission.

These three conditions were exactly those that applied to the Oklo deposits 2 billion years ago. The deposits were much larger than the minimum predicted size. Moreover, uranium-235 has a half-life of 704 million years, decaying about six times faster than the U-238 atoms, so several half-lives ago (around about 2 billion years) there would have been much more U-235 in the deposits—enough to lead to a sustainable chain reaction. Extrapolating backward, the relative proportions of the two isotopes would have been approximately 97 to 3, rather than 99.3 to 0.7 as it is today. And finally, the layers of rock had originally been in contact with groundwater, suggesting that what happened was the following:

A chain reaction would start in rocks surrounded by water, and the uranium atoms would split and generate heat. The heat would turn the water to steam, destroying its ability to moderate the reaction, and the neutrons would escape, stopping the reaction. The steam would condense back into water and blanket the neutron emission. More neutrons would be retained, splitting the atoms and restarting the chain reaction.

Explaining a tiny anomaly in the ratio of two types of atom in a small piece of rock has led to a description of a series of events that happened in a specific location on Earth billions of years ago. Over a period of 150 million years, a natural nuclear reactor would produce heat for about half an hour and then shut down for two and a half hours before starting up again, producing an average power of 100 kilowatts, like that produced in a typical car engine. Not only is this explanation deep, elegant, and beautiful, it's also incontrovertible. It doesn't depend on opinion or bias or desires, unlike many other "explanations" of how the world works, and that's the power of the best science.

KITTY GENOVESE AND GROUP APATHY

ADAM ALTER

Psychologist; assistant professor of marketing, Stern School of Business, New York University

The most elegant explanation in social psychology convinced me to pursue a PhD in the field. Every few years, a prominent tragedy attracts plenty of media attention because no one does anything to help. Just before sunrise on an April morning in 2010, a man lay dying on a sidewalk in Queens. The man, a homeless Guatemalan named Hugo Alfredo Tale-Yax, had intervened to help a woman whose male companion had begun shouting and shaking her violently. When Tale-Yax intervened, the man stabbed him several times in the torso. For ninety minutes, Tale-Yax lay in a growing pool of his own blood as dozens of passersby ignored him or stared briefly before continuing on their way. By the time emergency medical workers arrived to help, the sun had risen, and Tale-Yax was dead.

Almost half a century earlier, another New Yorker, Kitty Genovese, was attacked and ultimately killed while dozens of onlookers apparently failed to intervene. A *New York Times* writer decried the callousness of New Yorkers, and experts claimed that life in the city had rendered its citizens soulless. As they would in response to Tale-Yax's death, pundits wondered how dozens of people with functioning moral compasses could possibly have failed to help someone who was dying.

Social psychologists are taught to overcome the natural tendency to blame people for apparently bad behavior and to look instead for explanations in the environment. After Genovese's death, social psychologists Bibb Latané and John M. Darley were convinced that something about the situation explained the bystanders' failure to intervene. Their elegant insight was that human responses aren't additive in the same way that objects are additive. Whereas four lightbulbs illuminate a room more effectively than three lightbulbs, and three loudspeakers fill a room with noise more effectively than two loudspeakers, two people are often less effective than a single person. People second-guess situations, they stop to make sense of a chain of events before acting, and sometimes pride and the fear of looking foolish prevent them from acting at all.

In one of a series of brilliant studies, Latané and Darley videotaped students as they sat in a room that slowly filled with smoke.[*] The experimenters pumped smoke into the room with a smoke machine hidden behind a wall vent, the effect suggesting that there might be a fire nearby. When a subject was alone in the room, he usually left quickly and told the experimenter that something was amiss. But when a subject was surrounded by two or three others (some of them confederates, who were instructed to sit there immobile) he often remained seated, even as he lost sight of the others through the pall of smoke. When interviewed later, these students said they chose not to act because they concluded the smoke was benign (steam, or air-conditioning vapor), and they claimed that they had paid little or no attention to how the others in the room reacted.

According to Latané and Darley, the patterns of thinking that distinguish us from lower-order animals ultimately undermine our willingness to help in such situations, when we are alongside other people who are equally diffident.

---

[*] B. Latané & John M. Darley, "Group Inhibition of Bystander Intervention in Emergencies," *J. Pers. & Soc. Psych.* 10:3, 215–21 (1968).

THE WIZARD OF I

GERALD SMALLBERG

Practicing neurologist; playwright, Off-Off Broadway productions, Charter Members, The Gold Ring

Consciousness is the fusion of immediate stimuli with memory that combines the simultaneous feeling of being both the observer and the observed into a smooth, enveloping flow of time that is neither truly the past nor the present but somehow inexplicably each of them. It is the ultimate authority and arbiter of our perceptual reality. That consciousness is still an intractable problem for scientists and philosophers to understand is not surprising. Whatever the final answer turns out to be, I suspect it will be an illusion the mind evolved to hide the messy workings of its parallel modular computing.

Neurophysiologists are finding, as they pull ever so slightly at the veil that shrouds the "Wizard of I," that this indispensable attentive and observant self-monitor called consciousness is dependent on a trick in overseeing our perceptions. Our subjective sense of time does not correspond to reality. Cortical-evoked potentials—electrical recordings—of the normal brain during routine activity have been shown to precede by almost a third of a second the awareness of an actual willed movement or a response to sensory stimulation. The cortical-evoked potentials indicate that the brain is initiating or reacting to what is happening far sooner than the instantaneous perception we experience. On a physiological scale, this represents a huge discrepancy that our mind corrects by falsifying the actual time an action or event occurs, thus enabling our conscious experience to conform to what we perceive.

But data even more damaging to our confidence in the reliability of our perceptions come from studies of rapid eye movements, called saccades, that are triggered by novel visual stimuli. During the brief moments of these jerky eye movements, visual input to the brain is actively suppressed; we are literally blind. Without this involuntary ocular censorship, we would be repeatedly plagued with moments of acute blurred vision that would be unpleasant as well as unsafe. From a survival calculus, this would pose an extreme disadvantage, since it would invariably occur with novel stimuli, which by their very nature require not the worst but the best visual acuity.

The Wizard's solution to this intolerable situation is to exclude those intervals from our stream of awareness and replace them instead with a vision extrapolated from what just occurred to what is immediately anticipated. Consciousness, like a former president, has to come up with an accounting for the erased period. Evolution provided a much longer epic to work out the bugs in this necessary deception than the limited time frame under the gun of a special prosecutor. Instead of trying to hide the existence of the tape, consciousness came up with a far cleverer trick of obscuring the deletion. It does this by falsifying the time, backdating those necessary moments so that there is no appearance of any gap.

This illusion of visual continuity from inference and extrapolation reveals an innate vulnerability in the brain's software that any good hacker can exploit. Magicians, card sharks, three-card-monte hustlers have made a nice living working this perceptual flaw. In a comic routine, Richard Pryor expressed this best when caught by his wife with another woman. "Who are you going to believe? Me or your lying eyes?"

ONE COINCIDENCE; TWO DÉJÀ VUS

DOUGLAS COUPLAND
Writer, artist, designer; author, Marshall McLuhan:
You Know Nothing of My Work!

I take comfort in the fact that there are two human moments that seem to be doled out equally and democratically within the human condition, and that there is no satisfying ultimate explanation for either. One is coincidence, the other is déjà vu. It doesn't matter if you're Queen Elizabeth, one of the thirty-three miners rescued in Chile, a South Korean housewife, or a migrant herder in Zimbabwe: In the span of 365 days, you will pretty much have two déjà vus and one coincidence that makes you stop and say, "Wow, *that* was a coincidence!"

The thing about coincidence is that when you imagine the umpteen trillions of coincidences that can happen at any given moment, the fact is that in practice, coincidences almost never occur. They're so rare that when they do happen they're memorable. This suggests to me that the universe is designed to ward off coincidence whenever possible. The universe hates coincidence; I don't know why—it just seems to be true. So when a coincidence happens, that coincidence had to work awfully hard to escape the system. There's a message there. What is it? *Look.* Look *harder.* Mathematicians perhaps have a theorem for this, and if they do, it might, by default, be a theorem for something larger than what they think it is.

What's both eerie and interesting to me about déjà vus is that they occur almost metronomically throughout our lives—about once every six months, a poetic timekeeping device that, at the very least, reminds us we are alive. I can safely assume that my thirteen-year-old niece, Stephen Hawking, and someone working in a Beijing luggage factory each experience two déjà vus a year. Not one. Not three. *Two.*

The underlying biodynamics of déjà vu is probably ascribable to some sort of tingling neurons in a certain part of the brain, yet this doesn't tell us why they exist. They seem to me to be a signal from a larger point of view that wants to remind us that our lives are distinct, that they have meaning, and that they occur throughout a span of time. We are important, and what makes us valuable to the universe is our sentience and our curse and blessing of perpetual self-awareness.

OCCAM'S RAZOR

KATINKA MATSON
*Artist; cofounder, Edge.org; president,
Brockman, Inc.; author,* Short Lives

~~Keep it simple.~~

DEEP TIME

ALUN ANDERSON

Senior consultant and former editor-in-chief & publishing director, New Scientist; *author,* After the Ice: Life, Death, and Geopolitics in the New Arctic

There is one simple and powerful idea that strikes me as both deep and beautiful in its own right and as the mother of a suite of further elegant theories and explanations. The idea is that of "deep time"—that the Earth is extremely old and the life of our species on it has been extremely short. When that idea first emerged, it stood against everything that was then believed, and it was to eventually change people's view of themselves as much as did the earlier discovery that the Earth revolves around the sun.

We know when the idea of deep time was born, or at least first vindicated, because a University of Edinburgh professor named John Playfair recorded his reaction in 1788. "The mind seemed to grow giddy," he wrote, "by looking so far into the abyss of time."[*] He had traveled to the Scottish coast with his geologist friend James Hutton, who later put his ideas together in a book called *Theory of the Earth.* Hutton was showing him a set of distinct patterns in the rocks that could be most simply explained by assuming that the present land had been laid down in the sea, then lifted, distorted, eroded, and once again covered by new seafloor sediments. The Earth was not 6,000 years old, as then-accepted Biblical calculations decreed; nor had the strata precipitated out in a vast flood, as prevailing scientific views, informed by the best chemistry of the time, said.

---

[*]   *Trans. Roy. Soc. Edinburgh*, V, pt. III (1805).

It was an enormous shift to see the world as Hutton did. Appreciating the vastness of space is easy. When we look up at the stars, the immensity of the universe is both obvious and awe-inspiring. The immensity of time does not lie within human experience. Nature, observed on a human scale, passes only through the repeated cycle of the seasons, interrupted by occasional catastrophic earthquakes, volcanic eruptions, and floods. It is for that reason that creationist and catastrophic theories of Earth's origins appeared more plausible than those that were slow and gradual. But Hutton had faith in what he saw in the rocks, exhorting others to "open the book of Nature and read in her records."

His thinking about time created fertile ground for other grand theories. With huge spans of time available, imperceptibly slow processes could shape the natural world. After Hutton came modern geology, then the theory of evolution to explain how new species slowly arose, and eventually a theory of the gradual movement of the continents themselves. All are grounded in deep time.

Hutton's views were a huge challenge to religious orthodoxy, too, for when he wrote at the close of his book that "we find no vestige of a beginning—no prospect of an end," he challenged the idea of both a creation and a judgment day.

The beauty of his idea remains. If we look into the abyss of time we may not grow giddy, but we can feel our own insignificance in the Earth's 4.6-billion-year history and the significance of the precise moment in this vast span of time in which we live.

PLACING PSYCHOTHERAPY ON A SCIENTIFIC BASIS: FIVE EASY LESSONS

ERIC R. KANDEL

University professor and Kavli Professor of Brain Science,
Columbia University; author, The Age of Insight: The
Quest to Understand the Unconscious in Art, Mind,
and Brain, from Vienna 1900 to the Present

How did psychoanalysis, once a major mode for treating non-psychotic mental disorders, fall so badly in the estimation both of the medical community in the United States and the public at large? How can that be reversed? Let me try to address this question, putting it in some historical perspective.

While an undergraduate at Harvard College, I was drawn to psychiatry—and specifically to psychoanalysis. During my training, 1960–1965, psychotherapy was the major mode of treating mental illness, and this therapy was derived from psychoanalysis and based on the belief that one needed to understand mental symptoms in terms of their historical roots in childhood. These therapies tended to take years, and neither the outcome nor the mechanisms were studied systematically, because this was thought to be very difficult. Psychotherapy and psychoanalysis, when successful, allowed people to work a bit better and to love a bit, and these were dimensions that were considered difficult to measure.

In the 1960s, Aaron Beck changed all that by introducing several major obvious but nevertheless elegant and beautiful innovations:

First, he introduced instruments for measuring mental illness. Until Beck's work, psychiatric research was hampered by a dearth of techniques for operationalizing the various disorders and measuring their severity. Beck developed a number of instruments, beginning with a depression inventory, a hopelessness scale, and a suicide-intent scale. These scales helped objectify research in psychopathology and establish better clinical-outcome trials.

Second, Beck introduced a new short-term, evidence-based therapy he called cognitive behavioral therapy.

Third, Beck manualized the treatments; he wrote a cookbook, so method could be reliably taught to others. You and I could in principle learn to do cognitive behavioral therapy.

Fourth, he carried out, with the help of several colleagues, progressively better-controlled studies showing that cognitive behavioral therapy worked more effectively than placebos, and as effectively as antidepressants, in mild and moderate depression. In severe depression, it did not act as effectively as an antidepressant but acted synergistically with antidepressants to enhance recovery.

Beck's work was picked up by Helen Mayberg, another of my heroes in psychiatry. She carried out fMRI studies of depressed patients and discovered that Brodmann area 25 was a focus of abnormal activity in depression. She went on to find that if—and only if—a patient responded to cognitive behavior therapy or to antidepressant SSRIs (selective serotonin reuptake inhibitors), this abnormality reverted to normal.

What I find so interesting in this recital is the *Edge* Question: What elegant, deep explanation did Aaron Beck bring to his work that differentiated him from the rest of my generation of psychotherapists and allowed him to be so original?

Beck trained as a psychoanalyst in Philadelphia but soon became impressed with the radical idea that the central issue in many psychiatric disorders is not unconscious conflict but distorted patterns of thinking. He conceived this novel idea by listening with a critical—and open—mind to his patients with depression. In his early work on depression, Beck set out to test a specific psychoanalytic idea: that depression was due to "introjected anger." Patients with depression, it was argued, experienced deep hostility and anger toward someone they loved. They could not deal with having hostile feelings toward someone they valued, so they would repress their anger and direct it inward, toward themselves. Beck tested this idea by comparing the dreams—the royal road to the unconscious—of depressed patients with those of non-depressed patients and found that in their dreams depressed patients showed, if anything, less hostility than non-depressed patients. Instead, in their dreams, as in their waking lives, depressed patients have a systematic negative bias in their cognitive style, in the way they think about themselves and their future. They see themselves as "losers."

Beck considered these distorted patterns of thinking not simply a symptom—a reflection of a conflict lying deep within the psyche—but a key etiological agent in maintaining the disorders. This led him to develop a systematic psychological treatment for depression that focused on distorted thinking. He found that increasing the patients' objectivity regarding their misinterpretation of situations, or their cognitive distortions and negative expectations, resulted in substantial shifts in their thinking and subsequent improvements in their affect and behavior.

During his work on depression, Beck focused on suicide and provided for the first time a rational basis for the classification and assessment of suicidal behaviors, making it possible to identify high-risk individuals. His prospective study of 9,000 patients led to the formulation of an algorithm for predicting future suicide that has proved to have high predictive power. Of particular importance was his identification of clinical and psychological variables such as hopelessness and helplessness to predict future suicides. These turned out to be better predictors of suicide than clinical depression *per se*. Beck's work on suicide—and that of others, such as John Mann at Columbia—demonstrated that a short-term cognitive intervention can significantly reduce subsequent suicide attempts.

In the 1970s, Beck carried out the aforementioned controlled trials. Later, the National Institute of Mental Health carried out similar trials, and together these established cognitive therapy as the first psychological treatment shown to be effective in clinical depression.

As soon as cognitive therapy was found effective in the treatment of depression, Beck turned to other disorders. In a number of controlled clinical trials, he demonstrated that cognitive therapy is effective in panic disorder, post-traumatic-stress disorder, and obsessive-compulsive disorder. In fact, even before Helen Mayberg's work on depression, Lewis Baxter at UCLA had imaged patients with obsessive-compulsive disorder and found an abnormality in the caudate nucleus that was reversed when they improved with cognitive behavioral therapy.

Aaron Beck has recently turned his attention to patients with schizophrenia and has found that cognitive therapy helps improve their cognitive and negative symptoms, particularly their motivational deficits. Another amazing advance.

So the answer to the decline of psychoanalysis may not simply lie in the limitation of Freud's thought but much more in the lack of a deep, critical, scientific attitude of many in the subsequent generation of therapists. I have little doubt that insight therapy is extremely useful as a therapy, and there are studies supporting that contention. But an elegant, deep, and beautiful proof requires putting a set of highly validated approaches together to make the point in a convincing manner and perhaps even providing us with an idea of how the therapeutic result is achieved.

TRANSITIONAL OBJECTS

SHERRY TURKLE
*Abby Rockefeller Mauzé Professor of the Social Studies of Science
and Technology, MIT; author,* Alone Together: Why We Expect
More from Technology and Less from Each Other

I was a student in psychology in the mid-1970s at Harvard University. The grand experiment that had been "Social Relations" at Harvard had just crumbled. Its ambition had been to bring together the social sciences into one department—indeed, most of them into one building, William James Hall. Clinical psychology, experimental psychology, physical and cultural anthropology, and sociology—all of these would be in close quarters and intense conversation.

But now everyone was back in their own department, on their own floor. From my point of view, what was most difficult was that the people who studied thinking were on one floor and the people who studied feeling were on another.

In this Balkanized world, I took a course with George Goethals in which we learned about the passion in thought and the logical structure behind passion. Goethals, a psychologist who specialized in adolescence, was teaching a graduate seminar in psychoanalysis. His focus was on a particular school of analytic thought: British object-relations theory. This psychoanalytic tradition kept its eye on a deceptively simple question: How do we bring people and what they mean to us "inside" us? How do these internalizations cause us to grow and change? The "objects" of the theory's name were in fact people.

Several classes were devoted to the work of David Winnicott and his notion of the transitional object. Winnicott called "transitional" the objects of childhood—the stuffed animals, the bits of silk from a baby blanket, the favorite pillow—that the child experiences as both part of the self and of external reality. Winnicott writes that such objects mediate between the child's sense of connection to the body of the mother and a growing recognition that he or she is a separate being. The transitional objects of the nursery are destined to be abandoned; yet, says Winnicott, they leave traces that will mark the rest of life. Specifically, they influence how easily an individual develops a capacity for joy, aesthetic experience, and creative playfulness. Transitional objects, with their joint allegiance to self and other, demonstrate to the child that objects in the external world can be loved.

Winnicott believes that during all stages of life we continue to search for objects we experience as both within and outside the self. We give up the baby blanket, but we continue to search for the feeling of oneness it provided. We find this in moments of feeling "at one" with the world—what Freud called the "oceanic feeling." We experience these moments when we are at one with a piece of art, a vista in nature, a sexual experience.

As a scientific proposition, the theory of the transitional object has its limitations. But as a way of thinking about connection, it provides a powerful tool for thought. Specifically, it offered me a way to begin to understand the new relationships people were beginning to form with computers—something I began studying in the late 1970s and early 1980s. From the beginning, as I began to study the nascent digital culture, I could see that computers were not "just tools." They were intimate machines. People experienced them as part of the self, separate but connected to the self.

A novelist using a word-processing program referred to "my ESP with the machine. The words float out. I share the screen with my words." An architect who used the computer to design went even further: "I don't see the building in my mind until I start to play with shapes and forms on the machine. It comes to life in the space between my eyes and the screen." After studying programming, a thirteen-year-old girl said that when working with the computer, "there's a little piece of your mind, and now it's a little piece of the computer's mind, and you come to see yourself differently." A programmer talked about his "Vulcan mind meld" with the computer.

When I began studying the computer's special evocative power, my time with George Goethals and the small circle of Harvard graduate students immersed in Winnicott came back to me. Computers serve as transitional objects. They bring us back to the feelings of being "at one" with the world. Musicians often hear music in their minds before they play it, experiencing the music from within and without. The computer, similarly, can be experienced as an object on the border between self and not-self. Just as musical instruments can be extensions of the mind's construction of sound, computers can be extensions of the mind's construction of thought.

This way of thinking about the computer, as an evocative object, puts us on the inside of a new inside joke. For whenever psychoanalysts talked about object relations, they were always talking about people. From the beginning, people saw computers as "almost alive" or "sort of alive." With the computer, object-relations psychoanalysis can now be applied to, well, objects. People feel at one with video games, with lines of computer code, with the avatars they play in virtual worlds, with their smartphones. Classical transitional objects are meant to be abandoned, their power recovered in moments of heightened experience. When our current digital devices—our smartphones and cell phones—take on the power of transitional objects, a new psychology comes into play. These digital objects are never meant to be abandoned. We are meant to become cyborgs.

NATURAL SELECTION IS SIMPLE BUT THE SYSTEMS IT SHAPES ARE UNIMAGINABLY COMPLEX

RANDOLPH NESSE

*Professor of psychiatry and psychology, University of Michigan;
coauthor (with George C. Williams),* Why We Get Sick:
The New Science of Darwinian Medicine

The principle of natural selection is exceedingly simple. If some individuals in a population have a heritable trait associated with having more offspring, that trait will usually become more common in the population over the generations.

The products of natural selection are vastly complex. They are not merely complicated in the way that machines are complicated; they are organically complex in ways that are fundamentally different from any product of design. This makes them difficult for human minds to fully describe or comprehend. So we use that grand human gambit for understanding, a metaphor—in this case, the body as machine.

This metaphor makes it easy to portray the systems that mediate cell division, immune responses, glucose regulation, and all the rest, using boxes for the parts and arrows to indicate what causes what. Such diagrams summarize important information in ways we can grasp. Teachers teach them. Students dutifully memorize them. But they fundamentally misrepresent the nature of organic complexity. As John Scott Haldane noted in a prescient 1917 book, "a living organism has, in truth, but little resemblance to an ordinary machine."[*] Machines are designed; they have discrete parts with specific functions, and most remain intact when turned off. Individual machines are manufactured following identical copies of a single blueprint. In contrast, organisms evolve. They have components with indistinct boundaries and multiple functions that interact with myriad other parts and the environment to create self-sustaining reproducing systems whose survival requires the constant activity and cooperation of thousands of interdependent subsystems. Individual organisms develop from unique combinations of genes interacting with one another and with environments to create phenotypes, no two of which are identical.

Thinking about the body as a machine was a grand advance in the 16th century, when it offered an alternative to vitalism and vague notions of the life force. Now it's outmoded. It distorts our view of biological systems by fostering a tendency to think of them as simpler and more sensibly "designed" than they are. Experts know better. They recognize that the mechanisms regulating blood clotting are represented only crudely by the neat diagrams medical students memorize; most molecules in the clotting system interact with many others. Experts on the amygdala know that it has many functions, not just one or two, and they are mediated by scores of pathways to other brain loci. Serotonin exists not mainly to regulate mood and anxiety; it is essential to vascular tone, intestinal motility, and bone deposition. Leptin is not mainly a fat hormone; it has many functions, performing different ones at different times, even in the same cell. The reality of organic systems is vastly untidy. If only their parts were all distinct, with specific functions for each! Alas, these systems are not like machines. Our human minds have as little intuitive feeling for organic complexity as they do for quantum physics.

---

* John Scott Haldane, *Organism and Environment as Illustrated by the Physiology of Breathing* (New Haven, CT: Yale University Press, 1917), p. 91.

Recent progress in genetics confronts the problem. Naming genes according to postulated functions is as natural as defining chairs and boats by their functions. If each gene were a box on a blueprint labeled with its specific function, biology would be so much more tractable! However, it is increasingly clear that most traits are influenced by many genes, and most genes influence many traits. For instance, about 80 percent of the variation in human height is accounted for by genetic variation. It would seem straightforward to find the responsible genes. But looking for them has revealed that the 180 loci with the largest effects together account for only about 10 percent of the phenotypic variation. Recent findings in medical genetics are more discouraging. Just a decade ago, hope was high that we would soon find the variations that account for highly heritable diseases, such as schizophrenia and autism. But scanning the entire genome has revealed that there are no common alleles with large effects on these diseases. Some say we should have known. Natural selection would, after all, tend to eliminate alleles that cause disease. But thinking about the body as a machine aroused unrealistic hopes.

The grand vision for some neuroscientists is to trace every molecule and pathway to characterize all circuits in order to understand how the brain works. Molecules, loci, and pathways do serve differentiated functions; this is real knowledge, with great importance for human health. But understanding how the brain works by drawing a diagram that describes all the components and their connections and functions is a dream that may be unfulfillable. The problem is not merely fitting a million items on a page; the problem is that no such diagram can adequately describe the structure of organic systems. They are products of minuscule changes (from diverse mutations, migration, drift, and selection), which develop into systems with incompletely differentiated parts and incomprehensible interconnections—systems that nonetheless work very well indeed. Trying to reverse-engineer brain systems focuses important attention on functional significance, but it is inherently limited, because brain systems were never engineered in the first place.

Natural selection shapes systems whose complexity is indescribable in terms satisfying to human minds. Some may feel this is nihilistic. It does discourage hopes for finding simple specific descriptions for all biological systems. However, recognizing a quest as hopeless is often the key to progress. As Haldane put it, "We are thus brought face to face with a conclusion which to the biologist is just as significant and fundamental, and just as true to the facts observed, as the conclusion that mass persists is to the physicists. . . . [T]he structure of a living organism has no real resemblance in its behaviour to that of a machine. . . . In the living organism, . . . the 'structure' is only the appearance given by what seems at first to be a constant flow of specific material, beginning and ending in the environment."[*]

If bodies are not like machines, what are they like? They are more like Darwin's "tangled bank," with its "elaborately constructed forms, so different from each other, and dependent upon each other in so complex a manner."[†] Lovely. But can an ecological metaphor replace the metaphor of body as machine? Not likely. Perhaps someday understanding how natural selection shapes organic complexity will be so widely and deeply understood that scientists will be able to say, "A body is like . . . a living body," and everyone will know exactly what that means.

---

[*] Haldane, *Organization and Environment*, p. 99.

[†] Charles Darwin, *On the Origin of Species* (London: John Murray, 1872), p. 429.

HOW TO HAVE A GOOD IDEA

MARCEL KINSBOURNE

Professor of psychology, The New School; coauthor (with Paula J. Kaplan), Children's Learning and Attention Problems

You don't have to be human to have a good idea. You can even be a fish.

There's a large fish in shallow Micronesian waters that feeds on little fishes. The little fishes dwell in holes in the mud but swarm out to feed. The big fish starts to gobble up the little fishes, one by one, but they immediately retreat back into their holes when his meal has barely begun. What to do?

I have put this problem to my classes over the years, and I remember only one student who came up with the big fish's Good Idea. Of course he did it after a little thought, not after millions of years of evolution, but who's counting?

Here's the elegant trick. When the school of little fish appears, instead of gobbling, he swims low so that his belly rubs over the mud and blocks the escape holes. Then he can dine at leisure.

What do we learn? To have a good idea, stop having a bad one. The trick was to inhibit the easy, obvious, but ineffective attempts, permitting a better solution to come to mind. That worked for the big fish by some mechanism of mutation and natural selection in fish antiquity. Instead of tinkering with the obvious—obsessing about eating faster, taking bigger bites, etc.—junk plan A, and plan B comes swimming up. For humans, if the second solution doesn't work either, block that too, and wait. A third floats into awareness, and so on, until the insoluble is solved, even if the most intuitively obvious premises have to be overridden in the process.

To the novice, the Good Idea seems magical, a leap of intellectual lightning. More likely, however, it resulted from an iterative process as outlined above, with enough experience to help reject seductive but misleading premises. Thus the extraordinary actually arises, step by step, out of the ordinary.

Having a good idea is far from rare in the evolution of nonhuman species. Indeed, many if not most species need to have an idea or trick that works well enough for them to continue to exist. Admittedly, they may not be able to extrapolate its principle from the context in which it emerged and generalize it, as (some) people can, courtesy of their prefrontal cortex.

When the finest minds fail to resolve a classical problem over decades or centuries of trying, they were probably trapped by a premise that was so culturally "given" that it didn't even occur to them to challenge it—or they didn't notice it at all. But cultural context changes, and what seemed totally obvious yesterday becomes dubious, at best, today or tomorrow. Sooner or later, someone who may be no more gifted than his predecessors but is unshackled from some basic and incorrect assumption may hit upon the solution with relative ease.

Alternatively, one can be a fish, wait a million years or two, and see what comes up.

OUT OF THE MOUTHS OF BABES

NICHOLAS A. CHRISTAKIS

*Physician and social scientist, Harvard University; coauthor
(with James Fowler),* Connected: The Surprising Power of
Our Social Networks and How They Shape Our Lives

My favorite explanation is one that I sought as a boy. Why is the sky blue? It's a question every toddler asks, but it's also one that most great scientists since the time of Aristotle, including Leonardo da Vinci, Isaac Newton, Johannes Kepler, René Descartes, Leonhard Euler, and even Albert Einstein, have asked.

One of the things I like most about this explanation—beyond the artless simplicity of the question itself—is how many centuries of effort it took to arrive at and how many branches of science it involves.

Unlike other everyday phenomena, such as the rising and setting sun, the color of the sky did not elicit much myth-making, even by the Greeks or the Chinese. There were few nonscientific explanations for its color. It took a while for the azure sky to be problematized, but, when it was, it kept our (scientific) attention. How could the atmosphere be colored, when the air we breathe is not?

Aristotle is the first, as far as we know, to ask why the sky is blue. His answer, in the treatise *On Colors*, is that the air close at hand is clear and the deep air of the sky is blue in the same way that a thin layer of water is clear but a deep well of water looks black. This idea was still being echoed in the 13th century, by Roger Bacon. Kepler, too, reinvented a similar explanation, arguing that the air merely looks colorless because the tint of its color is so faint when in a thin layer. But none of them offered an explanation for the *blueness* of the atmosphere.

In the *Codex Leicester*, Leonardo da Vinci, writing in the early 16th century, noted, "I say that the blue which is seen in the atmosphere is not its own color, but is caused by the heated moisture having evaporated into the most minute, imperceptible particles, which beams of the solar rays attract and cause to seem luminous against the deep, intense darkness of the region of fire that forms a covering above them." Alas, Leonardo does not actually say why these particles should be blue either.

Newton contributed, both by asking why the sky was blue and by demonstrating, through his pathbreaking experiments with refraction, that white light could be decomposed into its constituent colors.

Many now-forgotten and many still-remembered scientists since Newton joined in. What might refract more blue light toward our eyes? In 1760, the mathematician Leonhard Euler speculated that the wave theory of light might help to explain why the sky is blue. The 19th century saw a flurry of experiments and observations of all sorts, from expeditions to mountaintops for observation to elaborate efforts to re-create the blue sky in a bottle—as chronicled in Peter Pesic's wonderful book, *Sky in a Bottle*. Countless careful observations of blueness at different locations, altitudes, and times were made, including with bespoke devices known as cyanometers. Horace-Bénédict de Saussure invented the first cyanometer in 1789. His version had fifty-three sections with varying shades of blue arranged in a circle. Saussure reasoned that something suspended in the air must be responsible for the blue color.

Indeed, for a very long time it had been suspected that something in the air modified the light and made it appear blue. Eventually it was realized that it was *the air itself* that did this—that the gaseous molecules that compose the air are essential to making it appear blue. And so the blueness of the sky is connected to the discovery of the physical reality of atoms. The color of the sky is deeply connected to atomic theory, and even to Avogadro's number. This in turn attracted Einstein's attention in the period from 1905 to 1910.

So, the sky is blue because the incident light interacts with the gas molecules in the air in such a fashion that more of the light in the blue part of the spectrum is scattered, reaching our eyes on the surface of the planet. All the frequencies of the incident light can be scattered this way, but the high-frequency (short wavelength) blue is scattered more than the lower frequencies in a process known as Rayleigh scattering, described in the 1870s. John William Strutt, Lord Rayleigh, who received the Nobel Prize in physics in 1904 for the discovery of argon, demonstrated that when the wavelength of the light is on the same order as the size of the gas molecules, the intensity of scattered light varies inversely with the fourth power of its wavelength. Shorter wavelengths like blue (and violet) are scattered more than longer ones. It's as if all the molecules in the air preferentially glowed blue, which is what we then see everywhere around us.

Yet the sky should appear violet, since violet light is scattered even more than blue light. The sky does not appear violet to us because of the final, biological part of the puzzle, which is the way our eyes are designed: They are more sensitive to blue than violet light.

The explanation for why the sky is blue involves much of the natural sciences: the colors in the visual spectrum, the wave nature of light, the angle at which sunlight hits the atmosphere, the mathematics of scattering, the size of nitrogen and oxygen molecules, and even the way human eyes perceive color. It's most of science, in a question a young child can ask.

THE BEAUTY IN A SUNRISE

PHILIP CAMPBELL
Editor-in-chief, Nature

When I was first excited by physics, the depths of its explanations were compelling to me in very esoteric contexts. For example, the binding of matter, energy, and spacetime in general relativity seemed (and indeed is) an extraordinarily elegant and deep explanation.

Nowadays I am even more compelled by powerful explanations that lie behind the things we see around us that are too easily taken for granted. And I find myself drawn to a context experienced every day by just about everybody.

> How generous is that himself the sun
> —arriving truly, faithfully who goes
> (never for a moment ceasing to begin
> the mystery of day for someone's eyes)

Thus wrote e. e. cummings in the opening of his lyrical celebration of our star. Those words highlight a daily moment—a sunrise—whose associated human sense of (in)significance and mystery may for some be only deepened by appreciating (at least) three great explanations underlying the experience. And each has at least one of the *Edge* Question's requested qualities: depth, elegance, and beauty.

If you care about such things, and (like me) live at a northern middle latitude, you will know the range of the horizon visible from your home, between roughly southeast and northeast, across which the point of sunrise shifts back and forth over the year, with sunrise getting later as it moves northward and the days shorten, and the motion reversing at the winter solstice. And beyond that quite complex behavior is the simple truth of the sun's fidelity—we can indeed trust it to come up somewhere in the east every morning.

Like a great work of art, a great scientific explanation loses none of its power to inspire awe afresh whenever one contemplates it. So it is with the explanation that those daily and annual cycles of sunrises are explicable by a tilted rotating Earth orbiting the sun, whose average axial direction can be considered fixed relative to the stars thanks to a still-mysterious conservation law.

Unlike my two other chosen explanations, this one encountered skepticism from scientists for decades. The heliocentric view of the solar system, articulated by Copernicus in the mid 16th century, was not widely accepted until well into the 17th. For me, that triumph over the combination of scientific skepticism and religious hostility only adds to the explanation's appeal.

Another explanation is certainly elegant and lies behind the changing hues of the sky as the sun rises. Lord Rayleigh succeeded James Clerk Maxwell as Cavendish professor of physics at Cambridge. One of his early achievements was to deduce laws of the scattering of light. His first effort reached the right answer on an invalid foundation—the scattering of light in an elastic aether. Although the existence of such an aether wasn't shown to be a fallacy until some years later, he redid his calculations using Maxwell's deeply unifying theories of electromagnetism. "Rayleigh scattering" is the expression of those theories in contexts where an electromagnetic wave encounters electrically polarized particles much smaller than its wavelength. The amount of scattering, Rayleigh discovered, is inversely proportional to the fourth power of the wavelength. By 1899, he had shown that air molecules themselves were powerful scatterers.

There, in one bound, is the essential explanation of why the sky is blue and why sunrises are reddened. Blue light is scattered much more by air molecules than light of longer wavelengths. The sun's disk is accordingly reddened, and all the more so when seen through the long atmospheric path at sunrise and sunset. (To fully account for the effect, you also need to take into account the sun's spectrum and the visual responses of human eyes.) The pink clouds that can add so much to the beauty of a sunrise consist of comparatively large droplets that scatter the wavelengths of reddened sunlight more equally than air molecules—colorwise, what you see is what they get.

The third explanation behind a sunrise is conceptually and cosmologically the deepest. What is happening in the sun to generate its seemingly eternal light and heat? Understanding the nuclear reactions at the sun's core was just a part of an explanation that, thanks especially to Burbidge, Burbidge, Fowler, and Hoyle in 1957,* simultaneously allowed us to understand not only the light from many kinds of stars but also how almost all the naturally occurring chemical elements are produced throughout the universe: in chains of reactions occurring within stable and cataclysmically unstable cosmic balls of gas in their various stages of stellar evolution, driven by the shifting influences of all the fundamental forces of nature—gravity, electromagnetism, and the strong and weak nuclear forces.

Edge readers know that scientific understanding enhances rather than destroys nature's beauty. All of these explanations, for me, contribute to the beauty in a sunrise.

Ah, but what is the explanation of beauty? Brain scientists grapple with nuclear-magnetic-resonance images; a recent meta-analysis indicated that all of our aesthetic judgments seem to

---

* "Synthesis of the Elements in Stars," *Rev. Mod. Phys.* 29:4, 547–650 (1957).

include the use of neural circuits in the right anterior insula, an area of the cerebral cortex typically associated with visceral perception. Perhaps our sense of beauty is a by-product of the evolutionary maintenance of the senses of belonging and of disgust. For what it's worth, as exoplanets pour out of our telescopes, I believe we will encounter astrochemical evidence for some form of extraterrestrial organism well before we achieve a deep, elegant, or beautiful explanation of human aesthetics.

THE ORIGIN OF MONEY

DYLAN EVANS

Founder and CEO of Projection Point; author, Risk
Intelligence: How to Live with Uncertainty

Carl Menger's account of the origin of money is my favorite scientific explanation. It's deeply satisfying because it shows how money can develop from barter without anyone consciously inventing it. As such, it's a great example of Adam Smith's Invisible Hand, or what scientists now call "emergence."

Menger (1840–1921) founded the Austrian school of economics, a heterodox school of thought derided by many mainstream economists. Yet their accounts of the origin of money beg the very question Menger answered. The typical mainstream-economics textbook lists the problems of barter exchange and then explains how money overcomes these problems. However, that doesn't really explain how money actually got started, any more than listing the advantages of air travel explains how airplanes were invented. As Lawrence White puts it in *The Theory of Monetary Institutions* (1999), "[O]ne is left with the impression that barterers, one morning, suddenly became alert to the benefits of monetary exchange, and, by that afternoon, were busy using some good as money."

That, of course, is ridiculous. In Menger's account, money emerges through a series of small steps, each of which is based on self-interested choices by individual traders with limited knowledge. First, individual barterers realize that when direct exchange is difficult, they can get what they want by indirect exchange. Rather than finding someone who both has what I want and wants what I have, I need only find someone who wants what I have. I can then trade what I have for his good, even though I don't want to consume it myself, and then trade that for something I do want to consume. In that case, I will have used the intermediate good as a medium of exchange.

Menger notes that not all goods are equally marketable; some goods are easier to trade than others. It therefore pays a trader to accumulate an inventory of highly marketable items for use as media of exchange. Other alert traders in the market catch on, and eventually the market converges on a single common medium of exchange. This is money.

Menger's theory shows not only how money can evolve without any conscious plan, but also that it doesn't depend on legal decrees or central banks. Yet this, too, is often overlooked by mainstream economists. Take Michael Woodford, for example. Woodford is one of the most influential academic monetary economists alive today, yet in his 2003 book *Interest and Prices: Foundations of a Theory of Monetary Policy*, a central bank somehow becomes part of the economy within less than a page after the initial assumptions are introduced. Woodford does not even pause for a page to consider what a banking system would look like without a central bank. Yet free banking has a long history; the first such system began in China in about A.D. 995, more than 600 years before the first central bank.

Can we account for the emergence of central banking by the same Invisible Hand–type of explanation Menger proposed for money? The answer depends, according to Lawrence White, on just what we mean by the term "central banking." If government sponsorship is among the defining features of a central bank, then the answer is no. The emergence of central banks cannot be accounted for entirely by market forces; at some point, deliberate state action must enter the story. The government's motives for getting involved aren't hard to imagine. For one thing, an exclusive supply of banknotes gives the government a source of monopoly profits in the form of a zero-interest loan from the public's holding of these non-interest-bearing notes.

In these troubled times, when central banks are expanding the stock of high-powered money through massive quantitative easing, Menger's theory is more relevant than ever. It alerts us to the possibility that the response to the current crisis in the euro zone need not be ever more centralization but could instead consist of a move in the opposite direction—toward a regime in which any bank is free to issue its own banknotes, and market forces, not central banks, control the money supply.

THE PRECESSION OF THE SIMULACRA

DOUGLAS RUSHKOFF

Media analyst; documentary writer; author, Life Inc: How
Corporatism Conquered the World, and How We Can Take It Back

Having discovered much too late in life that the many things I had
taken for granted as preexisting conditions of the universe were,
in fact, creations and ideas of people, I found French sociologist
and philosopher Jean Baudrillard's "precession of the simulacra"
to be an immensely valuable way of understanding just how disconnected from anything to do with reality we can become.

The main idea is that there's the real world, there's the maps
we use to describe that world, and then there's all this other activity that occurs on the map—sometimes with little regard for the
territory it's supposed to represent. There's the real world, there's
the representation of the world, and there's the mistaking of this
simulation for reality.

This idea came back into vogue when virtual reality was hitting
the scene. Writers evoked Baudrillard as if we needed to be warned
about escaping into our virtual worlds and leaving the brick-and-
mortar, flesh-and-blood one behind. But I never saw computer
simulations as so very dangerous. If anything, the obvious fake-
ness of computer simulations—from arcade games to Facebook—
not only kept us aware of their simulated nature but called into
question the reality of everything else.

So there's the land—this real stuff we walk around on. Then
there's territory—the maps and lines we use to define the land. But
then there are wars fought over where those map lines are drawn.

The levels can keep building on one another, bringing people
to further abstractions and disconnection from the real world.

Land becomes territory; territory then becomes property that is owned. Property itself can be represented by a deed, and the deed can be mortgaged. The mortgage is itself an investment that can be bet against with a derivative, which can be secured with a credit default swap.

The computer algorithm trading credit default swaps (as well as the programmers trying to follow that algorithm's actions in order to devise competing algorithms)—this level of interaction is real. And financially speaking, it has more influence over who gets to live in your house than almost any other factor. A credit-default-swap crisis can bankrupt a nation as big as the United States without changing anything about the real land it refers to.

Or take money: There's the thing of value—the labor, the chicken, the shoe. Then there's the thing we use to represent that value—say, gold, grain receipts, or gold certificates. But once we're accustomed to using those receipts and notes as the equivalent of a thing with value, we can go one step further: the federal reserve note, or "fiat" currency, which has no connection to gold, grain, or the labor, the chickens, the shoes. Three main steps: There's value, the representation of value, and then the disconnection from what has value.

But that last disconnection is the important one—the sad one, in many respects. Because that's the moment we forget where things came from—when we forget what they represent. The simulation is put forth as reality. The invented landscape is naturalized and then mistaken for nature.

And that is when we become particularly vulnerable to illusion, abuse, and fantasy. Because once we're living in a world of created symbols and simulations, whoever has control of the map has control of our reality.

TIME PERSPECTIVE THEORY

PHILIP ZIMBARDO

Professor emeritus of psychology, Stanford University; author, The
Lucifer Effect: Understanding How Good People Turn Evil

I am here to tell you that the most powerful influence on our every
decision that can lead to significant action outcomes is something
most of us are totally unaware of and yet is the most obvious psychological concept imaginable.

I'm talking about our sense of psychological time—more specifically, how our decisions are framed by the time zones we have
learned to prefer and tend to overuse. We all live in multiple time
zones, learned from childhood, shaped by education, culture,
social class, and experiences with economic and family stability or
instability. Most of us develop a biased temporal orientation that
favors one time frame over others, becoming excessively oriented
to past, present, or future.

Thus, at decision time for major or minor judgments, some of us
are totally influenced by factors in the immediate situation: what
others are doing, saying, urging, and one's own biological urges.
Others facing the same decision matrix ignore all those present
qualities, focusing instead on the past: the similarities between
current and prior settings, remembering what was done and its
effects. A third set of decision makers ignores the present and the
past and focuses primarily on the future consequences of current
actions, calculating costs vs. gains.

To complicate matters, there are subdomains of each of these
primary time zones. Some past-oriented people tend to focus
on negatives in their earlier experiences (regret, failure, abuse,
trauma), while others are primarily past-positive, focusing instead

on the good old days, nostalgia, gratitude, and successes. There are two ways to be present-oriented: to live in a present-hedonistic domain (seeking pleasure, novelty, sensation) or being present-fatalistic, believing nothing you do can change your future. Future-oriented people are goal setters, who plan strategies, which tend to be successful; but another future focus is on the transcendental future—life begins after death of the mortal body.

My interest in time-perspective theory inspired me to create an inventory so as to determine exactly the extent to which we fit into each of these six time zones. The Zimbardo Time Perspective Inventory, or ZTPI, correlates scores on these time dimensions with a host of other psychological traits and behaviors. We have demonstrated that time perspective has a major effect in a vast domain of human nature. In fact, some of the relationships uncovered reveal correlation coefficients much greater than any seen in traditional personality assessments. For example, future orientation correlates .70 with the trait of conscientiousness, which in turn predicts longevity. Present hedonism correlates .70 with sensation-seeking and novelty-seeking. Those high on past-negative are most likely to be high on measures of anxiety, depression, and anger, with correlations as robust as .75. Similarly substantial correlations are uncovered between present fatalism and these measures of personal distress. I should add that this confirmatory factor analysis was conducted on a sample of functioning college students, thus such effects should be cause for alarm by counselors. Beyond mere correlations of scale measures, the ZTPI scales predict a wide range of behaviors: course grades, risk-taking, alcohol, drug use and abuse, environmental conservation, medical checkups, creativity, problem solving, and much more.

Finally, one of the most surprising discoveries is the application of time-perspective theory to time therapy in "curing" PTSD in veterans, sexually abused women, or people suffering from motor-vehicle fatality experiences. Drs. Richard and Rosemary Sword have been treating with remarkably positive outcomes a number of veterans from recent U.S. wars and also civilian clients. The core of the treatment replaces the past-negative and present-fatalistic biased time zones common to those suffering from PTSD with a balanced time perspective that highlights the critical role of the hope-filled future, adds some selected present hedonism, and introduces memories of a past-positive nature. In a sample of thirty PTSD vets of varying ages and ethnicities treated with time perspective therapy for relatively few sessions (compared to traditional cognitive behavioral therapies), dramatic positive changes were found for all PTSD standard assessments, as well as in life-changing social and professional relationships. It is rewarding to see many of our veterans, who have continued to suffer for decades from their combat-related severe traumas, discover a new life rich with opportunities, friends, family, fun, and work by being exposed to this simple, elegant reframing of their mental orientation.

DEVELOPMENTAL TIMING EXPLAINS THE WOES OF ADOLESCENCE

ALISON GOPNIK

Professor of psychology and affiliate professor of philosophy at the University of California–Berkeley; author, The Philosophical Baby

"What was he thinking?" This is the familiar bewildered cry of parents trying to explain why their teenaged children act the way they do. Developmental psychologists, neuroscientists, and clinicians have an interesting and elegant explanation for teenage weirdness. It applies to a wide range of adolescent behavior, from the surprisingly admirable to the mildly annoying to the downright pathological. The idea is that there are two different neural and functional systems that interact to turn children into adults. The developmental relationship between those two systems has changed, and that, in turn, has profoundly changed adolescence.

First, there is a motivational and emotional system that's very closely linked to the biological and chemical changes of puberty. This is what turns placid ten-year-olds, safe in the protected immaturity of childhood, into restless, exuberant, emotionally intense teenagers, desperate to attain every goal, fulfill every desire, and experience every sensation. And for adolescents, the most important goal of all is to get the respect of your peers. Recent studies show that adolescents aren't reckless because they underestimate risks but because they overestimate rewards, especially social rewards—or, rather, that they find rewards more rewarding than adults do. Think about the incomparable intensity of first love, the never-to-be-recaptured glory of the high-school basketball championship. In youth you want things, and then in middle age you want to want them.

The second system is a control system that can channel and harness all that seething energy. The prefrontal cortex reaches out to guide and control other parts of the brain. This is the system that inhibits impulses and guides decision making. This control system depends much more on learning than does the motivational system. You get to make better decisions by making not-so-good decisions and then correcting them. You get to be a good planner by making plans, implementing them, and seeing the results again and again. Expertise comes with experience.

In the distant evolutionary past—in fact, even in the recent historical past—these two systems were in sync. Most childhood education involved formal and informal apprenticeships. Children had lots of chances to practice exactly the skills they would need to accomplish their goals as adults and so to become expert planners and actors. To become a good gatherer or hunter, cook or caregiver, you would actually practice gathering, hunting, cooking, and taking care of children all through middle childhood and early adolescence—tuning up just the prefrontal wiring you'd need as an adult. But you'd do all that under expert adult supervision and in the protected world of childhood, where the impact of your inevitable failures would be blunted. When the motivational juice of puberty kicked in, you'd be ready to go after the real rewards with new intensity and exuberance, but you'd also have the skill and control to do it effectively and reasonably safely.

In contemporary life, though, the relationship between these two systems has changed. For reasons that are somewhat mysterious but most likely biological, puberty is kicking in at an earlier and earlier age. (The leading theory points to changes in energy balance as children eat more and move less.) The motivational system kicks in with it.

At the same time, contemporary children have very little experience with the kinds of tasks they'll have to perform as grown-ups. Children have less and less chance even to practice such basic skills as cooking and caregiving. In fact, contemporary adolescents and pre-adolescents often don't do much of anything except go to school. The experience of trying to achieve a real goal in real time in the real world is increasingly delayed, and the development of the control system depends on just those experiences. The developmental psychologist Ron Dahl has a nice metaphor for the result: Adolescents develop a gas pedal and accelerator a long time before they get steering and brakes.

This doesn't mean that adolescents are stupider than they used to be; in many ways, they're much smarter. In fact, there's some evidence that delayed frontal development is correlated with higher IQ. The increasing emphasis on schooling means that children know more about more subjects than they ever did in the days of apprenticeships. Becoming a really expert cook doesn't tell you about the evolution of tool use or the composition of sodium chloride—the sorts of things you learn in school. But there are different ways of being smart; knowing history and chemistry is no help with a soufflé. Wide-ranging, flexible, and broad-based learning may hinder the development of finely honed, controlled, focused expertise in a particular skill.

Of course, the old have always complained about the young. But this explanation does elegantly account for the paradoxes and problems of our particular crop of adolescents. There do seem to be many young adults who are enormously smart and knowledgeable but directionless, who are enthusiastic and exuberant but unable to commit to a particular work or a particular love until well into their twenties or thirties. And there is the graver case of children faced with the uncompromising reality of the drive for sex, power, and respect, without the expertise and impulse control it takes to ward off pregnancy or violence.

I like this explanation because it accounts for so many puzzling everyday phenomena. But I also like it because it emphasizes two important facts about minds and brains that are often overlooked. First, there's the fact that experience shapes the brain. It's truer to say that our experience of controlling our impulses makes the prefrontal cortex develop than it is to say that prefrontal development makes us better at controlling our impulses.

Second, it's increasingly apparent that development plays a crucial role in explaining human nature. The old evolutionary psychology picture was that a small set of genes—a "module"—was directly responsible for some particular pattern of adult behavior. But there's more and more evidence that genes are just the first step in complex developmental sequences, cascades of interactions between organism and environment, and that those developmental processes shape the adult brain. Even small changes in developmental timing can lead to big changes in who we become.

IMPLICATIONS OF IVAN PAVLOV'S GREAT DISCOVERY

STEPHEN M. KOSSLYN
Psychologist; director, Center for Advanced Study in the Behavioral Sciences, Stanford University

ROBIN ROSENBERG
Clinical psychologist; author, What's the Matter with Batman?

It's easy to imagine a politician's objecting to federal funds going to study how dogs drool. But failing to support such research would have been very shortsighted indeed. As part of his Nobel Prize–winning research on digestion, the great Russian physiologist Ivan Pavlov (1849–1936) measured the amount of saliva produced when dogs were given food. In the course of this work, he and his colleagues noticed something unexpected: The dogs began salivating well before they were fed. In fact, they salivated when they first heard the approaching footsteps of the person coming to feed them. That core observation led to the discovery of classical conditioning.

The key idea behind classical conditioning is that a neutral stimulus (such as the sound of approaching footsteps) comes to be associated with a stimulus (such as food) that reflexively produces a response (such as salivation)—and after a while, the neutral stimulus elicits the response produced reflexively by the paired stimulus. To be clear about the phenomenon, we'll need to take a few words to explain the jargon. The neutral stimulus becomes "conditioned," and hence is known as the conditioned stimulus (CS), whereas the stimulus that reflexively produces the response is known as the unconditioned stimulus (UCS). And the response produced by the UCS is called the unconditioned response (UR). Classical conditioning occurs when the CS is presented right before a UCS, so that after a while the CS by itself produces the response. When this occurs, the response is called a conditioned response (CR). In short, at the outset a UCS (such as food) produces a UCR (such as salivation); when a CS (the sound of the feeder's footsteps) is presented before the UCS, it soon comes to produce the response, a CR (salivation), by itself.

This simple process gives rise to a host of elegant and nonintuitive explanations.

For example, consider accidental deaths from drug overdoses. In general, narcotics users tend to take the drug in a specific setting, such as their bathroom. The setting initially is a neutral stimulus, but after someone takes narcotics in it a few times, the bathroom comes to function as a CS: As soon as the user enters the bathroom with narcotics, the user's body responds to the setting by preparing for the ingestion of the drug. Specific physiological reactions allow the body to cope with the drug, and those reactions become conditioned to the bathroom (in other words, the reactions become a CR). To get a sufficient high, the user must now take enough of the narcotic to overcome the body's preparation. But if the user takes the drug in a different setting, perhaps in a friend's bedroom during a party, the CR does not occur—that is, the usual physiological preparation for the narcotic does not take place. Thus, the usual amount of the drug functions as if it were a larger dose and may be more than the user can tolerate without the body's preemptive readiness. Hence, although the process of classical conditioning was formulated to explain very different phenomena, it can be extended to explain why drug overdoses sometimes accidentally occur when usual doses are taken in new settings.

By the same token, classical conditioning plays a role in the placebo effect: The analgesics regularly used by many of us, such as ibuprofen or aspirin, begin to take effect well before their active ingredients have time to kick in. Why? From previous experience, the mere act of taking that particular pill has become a CS, which triggers the pain-relieving processes invoked by the medicine itself (and those processes have become a CR).

Classical conditioning also can result from an implanted defibrillator, or pacemaker. When the heart beats too quickly, this device shocks it, causing it to revert to beating at a normal rate. Until the shock level is properly calibrated, the shock can be very uncomfortable and function as a UCS, producing fear as a UCR. Because the shock does not occur in a consistent environment, the person associates random aspects of the environment with it—which then function as CS's. And when any of those environmental aspects are present, the person can experience severe anxiety, awaiting the possible shock.

This same process explains why you find a particular food unappealing once it's given you food poisoning. It can thus come to function as a CS, and if you eat it—or even think about eating it—you may feel queasy, a CR. You may find yourself avoiding that food, and thus a food aversion is born. In fact, simply pairing pictures of particular types of food (such as French fries) with aversive photographs (such as of a horribly burned body) can change how appealing you find that food.

Thus Pavlov's discovery of anticipatory salivation can be easily extended to a wide range of phenomena. But that said, we should point out that his original conception of classical conditioning was not quite right. He thought that sensory input was directly connected to specific responses, leading the stimuli to produce the response automatically. We now know that the connection is not so direct; classical conditioning involves many cognitive processes, such as attention and those underlying interpretation and understanding. In fact, classical conditioning is a form of implicit learning. As such, it allows us to navigate through life with less cognitive effort (and stress) than would otherwise be required. Nevertheless, this sort of conditioning has by-products that can be powerful, surprising, and even sometimes dangerous.

NATURE IS CLEVERER THAN WE ARE

TERRENCE J. SEJNOWSKI
Computational neuroscientist; Francis Crick Professor, the Salk Institute; coauthor (with Patricia S. Churchland), The Computational Brain

We have the clear impression that our deliberative mind makes the most important decisions in our life: what work we do, where we live, whom we marry. But contrary to this belief, the biological evidence points toward a decision process in an ancient brain system called the basal ganglia, brain circuits that consciousness cannot access. Nonetheless, the mind dutifully makes up plausible explanations for the decisions.

The scientific trail that led to this conclusion began with honeybees. Worker bees forage the spring fields for nectar, which they identify with the color, fragrance, and shape of a flower. The learning circuit in the bee brain converges on VUMmx1, a single neuron that receives the sensory input and, a bit later, the value of the nectar, and learns to predict the nectar value of that flower the next time the bee encounters it. The delay is important, because the key is prediction, rather than a simple association. This is also the central core of temporal-difference (TD) learning, which entails learning a sequence of decisions leading to a goal and is particularly effective in uncertain environments, like the world we live in.

Buried deep in your midbrain, there's a small collection of neurons—found in our earliest vertebrate ancestors, and projecting throughout the cortical mantle and basal ganglia—that are important for decision making. These neurons release a neurotransmitter called dopamine, which has a powerful influence on our behavior. Dopamine has been called a "reward molecule," but more important than reward itself is the ability of these neurons to predict reward: If I had that job, how happy would I be? Dopamine neurons, which are central to motivation, implement TD learning, just as VUMmx1 does.

TD learning solves the problem of finding the shortest path to a goal. It's an on-line algorithm, because it learns by exploring and discovers the value of intermediate decisions in reaching the goal. It does this by creating an internal value function, which can be used to predict the consequences of actions. Dopamine neurons evaluate the current state of the entire cortex and inform the brain about the best course of action from the current state. In many cases, the best course of action is a guess, but because guesses can be improved, TD learning creates, over time, a value function of oracular powers. Dopamine may be the source of the "gut feeling" you sometime experience, the stuff that intuition is made from.

When you're considering various options, prospective brain circuits evaluate each scenario, and the transient level of dopamine registers the predicted value of each decision. The level of dopamine is also related to your level of motivation, so not only will a high level of dopamine indicate a high expected reward, but you will also have a higher level of motivation to pursue it. This is quite literally the case in the motor system, where a higher tonic dopamine level produces faster movements. The addictive power of cocaine and amphetamines is a consequence of increased dopamine activity, hijacking the brain's internal motivation system. Reduced levels of dopamine lead to anhedonia, an inability to experience pleasure; and the loss of dopamine neurons results in Parkinson's disease, an inability to initiate actions and thoughts.

TD learning is powerful because it combines information about value along many different dimensions, in effect comparing apples and oranges in achieving distant goals. This is important because rational decision making is very difficult when there are many variables and unknowns. Having an internal system that quickly delivers good guesses is a great advantage, and may make the difference between life and death when a quick decision is needed. TD learning depends on the sum of your life experiences. It extracts what is essential from these experiences long after the details of the individual experiences are no longer remembered.

TD learning also explains many of the experiments performed by psychologists who trained rats and pigeons in simple tasks. Reinforcement learning algorithms have traditionally been considered too weak to explain complex behaviors, because the feedback from the environment is minimal. Nonetheless, reinforcement learning is universal among nearly all species and is responsible for some of the most complex forms of sensorimotor coordination, such as piano playing and speech. Reinforcement learning has been honed by hundreds of millions of years of evolution. It has served countless species well, particularly our own.

How complex a problem can TD learning solve? TD-Gammon is a computer program that learned how to play backgammon by playing itself. The difficulty with this approach is that the reward comes only at the end of the game, so it's not clear which were the good moves that led to the win. TD-Gammon started out with no knowledge of the game, except for the rules. By playing itself many times and applying TD learning to create a value function to evaluate game positions, TD-Gammon climbed from beginner to expert level, along the way picking up subtle strategies similar to ones that humans use. After playing itself a million times, it reached championship level and was discovering new positional play that astonished human experts. Similar approaches to the game of Go have achieved impressive levels of performance and are on track to reach professional levels.

When there's a combinatorial explosion of possible outcomes, selective pruning is helpful. Attention and working memory allow us to focus on most of the important parts of a problem. Reinforcement learning is also supercharged by our declarative memory system, which tracks unique objects and events. When large brains evolved in primates, the increased memory capacity greatly enhanced their ability to make complex decisions, leading to longer sequences of actions to achieve goals. We are the only species to create an educational system and to consign ourselves to years of instruction and tests. Delayed gratification can extend into the distant future (in some cases, into an imagined afterlife), a tribute to the power of dopamine to control behavior.

At the beginning of the cognitive revolution in the 1960s, the brightest minds could not imagine that reinforcement learning could underlie intelligent behavior. Minds are not reliable. Nature is cleverer than we are.

IMPOSING RANDOMNESS

MICHAEL I. NORTON

Associate professor of business administration and
Marvin Bower Fellow, Harvard Business School

Paul Meier, who passed away in 2011, was primarily known for his introduction of the Kaplan-Meier estimator. But Meier was also a seminal figure in the widespread adoption of an invaluable explanatory tool: the randomized experiment. The decided unsexiness of the term masks a truly elegant form, which in the hands of its best practitioners approaches art. Simply put, experiments offer a unique and powerful means for devising answers to the question that scientists across disciplines seek to answer: How do we know whether something works?

Take a question that appears anew in the media each year: Is red wine good or bad for us? We learn a great deal about how red wine works by asking people about their consumption and health and looking for correlations between the two. To estimate the specific impact of red wine on health, though, we need to ask people a *lot* of questions—about everything they consume (food, prescription medication, more unsavory forms of medication), their habits (exercise, sleep, sexual activity), their past (their health history, their parents' and grandparents' health histories), and on and on—and then try to control for these factors to isolate the impact of wine on health. Think of the length of the survey.

Randomized experiments completely reengineer how we go about understanding how red wine works. We take it as a given that people vary in the manifold ways described above (and others), but we cope with this variance by randomly assigning people to either drink red wine or not. If people who eat doughnuts and never exercise are equally likely to be in the "wine treatment" or the "control treatment," then we can do a decent job of assessing the average impact of red wine over and above the likely impact of other factors. It sounds simple because, well, it is—but anytime a simple technique yields so much, "elegant" is a more apt description.

The rise of experiments in the social sciences beginning in the 1950s—including Meier's contributions—has exploded in recent years with the adoption of randomized experiments in fields ranging from medicine (testing interventions, like cognitive behavioral therapy) to political science (running voter-turnout experiments) to education (assigning kids to be paid for grades) to economics (encouraging savings behavior). The experimental method has also begun to filter into and impact public policy: President Obama appointed behavioral economist Cass Sunstein to head the Office of Information and Regulatory Affairs, and Prime Minister David Cameron instituted a Behavioural Insights Team.

Randomized experiments are by no means a perfect tool for explanation. Some important questions simply do not lend themselves to randomized experiments, and the method in the wrong hands can cause harm, as in the infamous Tuskegee syphilis experiment. But their increasingly widespread application speaks to their flexibility in informing us how things work and why they work that way.

THE UNIFICATION OF ELECTRICITY AND MAGNETISM

LAWRENCE M. KRAUSS
Physicist/cosmologist, Arizona State University;
author, A Universe from Nothing

No explanation I know of in recent scientific history is as beautiful or deep, or ultimately as elegant, as the 19th-century explanation of the remarkable connection between two familiar but seemingly distinct forces in nature—electricity and magnetism. It represents to me all that is best about science: It combined surprising empirical discoveries with a convoluted path to a remarkably simple and elegant mathematical framework, which explained far more than was ever bargained for and in the process produced the technology that powers modern civilization.

Strange experiments with jumping frogs and electric circuits eventually led to the serendipitous discovery, by the self-schooled yet greatest experimentalist of his time, Michael Faraday, of a strange connection between magnets and electric currents. By then, it was well known that a moving electric charge (or current) created a magnetic field around itself that could repel or attract other nearby magnets. What remained an open question was whether magnets could produce any electric force on charged objects. Faraday discovered, by accident, that when he turned a switch on or off to start or stop a current, creating a magnetic field that grew or decreased with time, during the periods when the magnetic field was changing, a force would suddenly arise in a nearby wire, moving the electric charges within it to create a current.

Faraday's law of induction, as it became known, not only is

responsible for the basic principle governing all electric genera-tors from Niagara Falls to nuclear power plants but also produced a theoretical conundrum that required the mind of the great-est theoretical physicist of his time, James Clerk Maxwell, to set things straight. Maxwell realized that Faraday's result implied that it was the changing magnetic field (a pictorial concept intro-duced by Faraday himself because he felt more comfortable with pictures than algebra) that produced an electric field that pushed the charges around the wire, thereby creating a current.

Achieving mathematical symmetry in the equations governing electric and magnetic fields then required that a changing electric field and not merely moving charges would produce a magnetic field. This not only produced a set of mathematically consistent equations every physics student knows (and some love) called Maxwell's equations, which can fit on a T-shirt, but it established the physical reality of what was otherwise a figment of Faraday's imagination: a field—that is, some quantity associated with every point in space and time.

Moreover, Maxwell realized that if a changing electric field produced a magnetic field, then a constantly changing electric field, such as occurs when you continuously jiggle a charge up and down, would produce a constantly changing magnetic field. That, in turn, would create a constantly changing electric field, which would create a constantly changing magnetic field, and so on. This field "disturbance" would move out from the original source (the jiggling charge) at a rate that Maxwell could calculate on the basis of his equations. The parameters in these equations came from experiment—from measuring the strength of the electric force between two known charges and the strength of the magnetic force between two known currents.

From these two fundamental properties of nature, Maxwell cal-culated the speed of the disturbance and found out that the speed was precisely the speed that light was measured to have! Thus he discovered that light is indeed a wave—but a wave of electric and magnetic fields that moves through space at a precise speed determined by two fundamental constants in nature. This laid the basis for Einstein to come along a generation or so later and demonstrate that the constant speed of light required a revision in our notions of space and time.

So, from jumping frogs and differential equations came one of the most beautiful unifications in all of physics: the unification of electricity and magnetism in a single theory of electromagnetism. Maxwell's theory explained the existence of that which allows us to observe the universe around us—namely, light. Its practical implications would produce the mechanisms that power modern civilization and the principles that govern essentially all modern electronic devices. And the nature of the theory itself produced a series of further puzzles that allowed Einstein to come up with new insights into space and time!

Not bad for a set of experiments whose worth was questioned by Gladstone (or by Queen Victoria, depending on which apocryphal story you buy), who came into Faraday's laboratory and wondered what all the fuss was about and what use all of this experimentation was. He (or she) was told either "Of what use is a newborn baby?" or, in my favorite version of the story, "Use? Why, one day this will be so useful you will tax us for it!" Beauty, elegance, depth, utility, adventure, and excitement! Science at its best!

FURRY RUBBER BANDS

NEIL GERSHENFELD
Director, Center for Bits and Atoms, MIT; author, Fab:
The Coming Revolution on Your Desktop—from
Personal Computers to Personal Fabrication

I learned electrodynamics at Swarthmore, from Professor Mark Heald and his concise text on an even more concise set of equations, Maxwell's. In four lines, just thirty-one characters (or fewer, with some notational tricks), Maxwell's equations unified what had appeared to be unrelated phenomena (the dynamics of electric and magnetic fields), predicted new experimental observations, and contained both theoretical advances to come (including the wave solution for light and special relativity) and technologies to come (including the fiberoptics, coaxial cables, and wireless signals that carry the Internet).

But the explanation I found memorable was not Maxwell's of electromagnetism, which is well known for its beauty and consequence. It was Heald's explanation that electric field lines behave like furry rubber bands: They want to be as short as possible (the rubber) but don't want to be near each other (the fur). This is an easily understood, qualitative description that has served me in good stead in device design. And it provides a deeper quantitative insight into the nature of Maxwell's equations: The local solution for the field geometry can be understood as solving a global optimization.

These sorts of scientific similarities that are predictive as well as descriptive help us reason about regimes our minds didn't evolve to operate in. Unifying forces is not an everyday occurrence, but explaining them can be. Recognizing that something is precisely like something else is a kind of object-oriented thinking that helps build bigger thoughts out of smaller ideas.

I understood Berry's phase for spinors by trying to rotate my hand while holding up a glass; I mastered NMR spin echoes by swinging my arms while I revolved; the alignment of semiconductor Fermi levels at a junction made sense when explained as filling buckets with water. Like furry rubber bands and electric fields, these relationships represent analogies between governing equations. Unlike words, they can be exact, providing explanations that connect unfamiliar formalism with familiar experience.

THE PRINCIPLE OF INERTIA

LEE SMOLIN

Physicist, Perimeter Institute; author, The Trouble with Physics, The Life of the Cosmos, *and* Three Roads to Quantum Gravity

My favorite explanation in science is the principle of inertia. It explains why we can't feel the Earth in motion. This principle was perhaps the most counterintuitive and revolutionary step taken in all of science. It was proposed by both Galileo and Descartes and has been the core of countless successful explanations in physics in the centuries since. The principle is the answer to a very simple question: *How would an object that is free (in the sense that no external influences or force affects its motion) move?*

To answer this question, we need a definition of motion. What does it mean for something to move? The modern conception is that motion has to be described relative to an observer.

Consider an object at rest relative to you—say, a cat sleeping on your lap—and imagine how it appears to move as seen by other observers. Depending on how the observer is moving, the cat can appear to have any sort of motion at all. If the observer spins relative to you, the cat will appear to spin to that observer. So to make sense of the question of how free objects move, we have to refer to a special class of observers. The answer to the question is the following:

There is a special class of observers, relative to whom all free objects appear either to be at rest or to move in a straight line at a constant speed.

I have just stated the principle of inertia.

The power of this principle is that it is completely general. Once a special observer sees a free object move in a straight line with constant speed, she will observe all other free objects to so move.

Furthermore, suppose you're a special observer. Any observer who moves in a straight line at a constant speed with respect to you will also see the free objects move at a constant speed in a straight line with respect to him. Special observers form a big class, one whose members are all moving with respect to one another. These special observers are called *inertial observers.*

An immediate and momentous consequence is that there is no absolute meaning to not moving. An object may be at rest with respect to one inertial observer, but other inertial observers will see it moving—always in a straight line at constant speed. This can be formulated as a principle:

There is no way, by observing objects in motion, to distinguish observers at rest from other inertial observers.

Any inertial observer can plausibly say that he is the one at rest and the others are moving. This is called Galileo's principle of relativity. It explains why the Earth can move without our experiencing the gross effects.

To appreciate how revolutionary this principle was, notice that physicists of the 16th century could disprove, by a simple observation, Copernicus's claim that the Earth moves around the sun: Just drop a ball from the top of a tower. If the Earth was rotating around its axis and orbiting the sun at the speeds Copernicus required, the ball would land far from the tower, instead of at its base. *QED*: The Earth is at rest.

But this proof assumes that motion is absolute, defined with respect to a special observer at rest, with respect to whom objects with no forces on them come to rest. By altering the definition of motion, Galileo could argue that this same experiment shows that the Earth might indeed be moving.

The principle of inertia was the core of the Scientific Revolution of the 17th century; moreover, it contained the seeds of revolutions to come. To see why, notice the qualifier in the statement of Galileo's principle of relativity: "by observing objects in motion." For many years, it was thought that someday we would make other kinds of observations that would determine which inertial observers are really moving and which are really at rest. Einstein constructed his special theory of relativity simply by removing this qualifier. His principle of relativity states:

There is no way to distinguish observers at rest from other inertial observers.

And there's more. A decade after special relativity, the principle of inertia was the seed for the next revolution—the discovery of general relativity. The principle was generalized by replacing "moving in a straight line with constant speed" to "moving along a geodesic in spacetime." A geodesic is the generalization of a straight line in a curved geometry—it's the shortest distance between two points. So now the principle of inertia reads:

There is a special class of observers, relative to whom all free objects appear to move along geodesics in spacetime. These are observers who are in free fall in a gravitational field.

And there is a consequent generalization:

There is no way to distinguish observers in free fall from one another.

This becomes Einstein's equivalence principle, the core of his general theory of relativity.

But is the principle of inertia really true? So far, it has been tested in circumstances where the energy of motion of a particle is as much as 11 orders of magnitude greater than its mass. This is pretty impressive, but there's still a lot of room for the principle of inertia to fail. Only experiment can tell us whether it or its failure will be the core of revolutions in science to come.

But whatever the outcome, it is the only explanation in science to have survived unscathed for so long, to have proved valid over such a range of scales, and to have sparked so many scientific revolutions.

SEEING IS BELIEVING: FROM PLACEBOS TO MOVIES IN OUR BRAIN

ERIC J. TOPOL

Gary and Mary West Chair of Innovative Medicine and
professor of translational genomics, Scripps Research Institute;
author, The Creative Destruction of Medicine

Our brain—with its 100 billion neurons and quadrillion synapses give or take a few billion here or there—is one of the most complex entities to demystify. And that may be a good thing, since we don't necessarily want others reading our minds, which would take the recent megatrend of transparency much too far.

But the use of functional magnetic resonance (fMRI) and positron emission tomography (PET) to image the brain and construct sophisticated activation maps validates the "Seeing is believing" aphorism for any skeptics. One of the longest controversies in medicine has been whether the placebo effect, a notoriously complex mind-body endproduct, has a genuine biological mechanism. That controversy now seems to be resolved with the recognition that the opioid-drug pathway—induced by drugs like morphine and oxycontin—shares the same brain-activation pattern as that of the administration of placebos for pain relief. And dopamine release from specific regions of the brain has been detected after administering a placebo to patients with Parkinson's disease. Indeed, the upgrading of the placebo effect to include discrete, distinguishable psychobiological mechanisms has now prompted consideration of placebo medications as therapeutics—Harvard University recently set up a dedicated institute called the Program in Placebo Studies and the Therapeutic Encounter.

The decoding of the placebo effect seems a step on the way to the more ambitious quest of mind-reading. In the summer of 2011, a group at the University of California–Berkeley produced, by reconstructing brain-imaging activation maps, a reasonable facsimile of the short YouTube movies shown to their experiment's subjects.* In fact, it was inspiring and downright scary to see the resemblance of the frame-by-frame comparisons of the movies and what was reconstructed from the brain imaging.

Couple this with the ongoing development of miniature portable MRIs and we may be on the way to watching our dreams in the morning on our iPad. Or, even more worrisome, revealing the movies in our brain to anyone interested in seeing them.

---

* Shinji Nishimoto et al., "Reconstructing Visual Experiences from Brain Activity Evoked by Natural Movies," *Curr. Biol.* 21:19, 1641-46 (2011).

THE DISCONTINUITY OF SCIENCE AND CULTURE

GERALD HOLTON

Mallinckrodt Professor of Physics and professor of the history of science, emeritus, Harvard University; coeditor (with Peter Galison and Silvan Schweber), Einstein for the 21st Century: His Legacy in Science, Art, and Modern Culture

From time to time, large sections of humanity find themselves, at short notice, in a different universe. Science, culture, and society have undergone a tectonic shift, for better or worse—the rise of a powerful religious or political leader, the Declaration of Independence, the end of slavery—or, on the other hand, the fall of Rome, the Great Plague, the World Wars.

So, too, in the world of art. Thus Virginia Woolf said famously, "On or about December 1910 human character changed," owing, in her view, to the explosive exhibition of Post–Impressionist canvases in London that year. And after the discovery of the nucleus was announced, Wassily Kandinsky wrote: "The collapse of the atom model was equivalent, in my soul, to the collapse of the whole world. Suddenly, the thickest walls fell . . . ," and he could turn to a new way of painting.

Each of such worldview-changing occurrences tends to be deeply puzzling or anguishing. They are sudden fissures in the familiar fabric of history that ask for explanations, with treatises published year after year, each hoping to provide an answer, seeking the cause of the dismay.

I will here focus on one such phenomenon.

In 1611, John Donne published his poem *The First Anniver-*

sary, containing the familiar lines "And new Philosophy calls all in doubt, / The Element of fire is quite put out;" and later, " . . . Is crumbled out againe to his Atomies / 'Tis all in pieces, all coherence gone; / All just supply, and all Relation." He and many others felt that the old order and unity had been displaced by relativism and discontinuity. The explanation for his anguish was an entirely unexpected event the year before: Galileo's discovery of the fact that the moon has mountains, that Jupiter has moons, that there are immensely more stars than had been known.

Of this happening and its consequent findings, the historian Marjorie Nicolson wrote: "We may perhaps date the beginning of modern thought from the night of January 7, 1610, when Galileo, by means of the instrument he developed [the telescope], thought he perceived new planets and new, expanded worlds."[*]

Indeed, by his work Galileo gave a deep and elegant explanation for the question of how our cosmos is arranged—no matter how painful this may have been to the Aristotelians and poets of his time. At last, the Copernican theory, formulated long ago, had more credibility. From this vast step forward, new science and new culture could be born.

---

[*] *Science and Imagination* (Ithaca, NY: Cornell University Press, 1956), 4.

HORMESIS IS REDUNDANCY

NASSIM NICHOLAS TALEB

Distinguished Professor of Risk Engineering,
NYU-Poly; author, The Black Swan

Nature is the master statistician and probabilist. It follows a certain logic based on layers of redundancies, as a central risk-management approach. Nature builds with extra spare parts (two kidneys) and extra capacity in many, many things (say lungs, neural system, arterial apparatus, etc.), while designs by humans tend to be spare and overoptimized, and have the opposite attribute of redundancy—that is, leverage; we have a historical track record of engaging in debt, which is the reverse of redundancy ($50,000 in extra cash in the bank or, better, under the mattress, is redundancy; owing the bank an equivalent amount is debt).

Now, remarkably, the mechanism called hormesis is a form of redundancy and statistically sophisticated in ways human science (so far) has failed us.

Hormesis is when a bit of a harmful substance, or stressor, in the right dose or with the right intensity, stimulates the organism and makes it better, stronger, healthier, and prepared for a stronger dose the next exposure. That's the reason we go to the gym, engage in intermittent fasting or caloric deprivation, or overcompensate for challenges by getting tougher. Hormesis lost some scientific respect, interest, and practice after the 1930s, partly because some people mistakenly associated it with the practice of homeopathy. The association was unfairly done, as the mechanisms are extremely different. Homeopathy relies on other principles, such as the one that minute, highly diluted parts of the agents of a disease (so small they can hardly be perceptible, hence cannot cause hormesis) could help medicate against the disease itself. It has shown little empirical backing and belongs today to alternative medicine, while hormesis, as an effect, has shown ample scientific evidence.

Now it turns out that the logics of redundancy and overcompensation are the same—as if nature had a simple, elegant, and uniform style in doing things. If I ingest, say, 15 milligrams of a poisonous substance, my body will get stronger, preparing for 20, or more. Stressing my bones (karate practice or carrying water on my head) will cause them to prepare for greater stress by getting denser and tougher. A system that overcompensates is necessarily in overshooting mode, building extra capacity and strength in anticipation for the possibility of a worse outcome, in response to information about the possibility of a hazard. This is a very sophisticated form of discovering probabilities via stressors. And of course such extra capacity or strength becomes useful—in itself—as opportunistic as it can be used to some benefit even in the absence of the hazard. Redundancy is an aggressive, not a defensive, approach to life.

Alas, our institutional risk-management methods are vastly different. Current practice is to look in the past for the worst-case scenario, called a "stress test," and adjust accordingly, never imagining that, just as the past experienced a large deviation that did not have a predecessor, such deviation might be insufficient. For instance, current systems take the worst historical recession, the worst war, the worst historical move in interest rates, the worst point in unemployment, etc., as an anchor for the worst future outcome. Many of us have been frustrated—very frustrated—by the method of stress testing in which people never go beyond what has happened before, and have even had to face the usual expression of naive empiricism ("Do you have evidence?") when suggesting that we need to consider worse.

And, of course, these systems don't do the recursive exercise in their mind to see the obvious—that the worst past event itself did not have a predecessor of equal magnitude, and that someone using the past worst case in Europe before the Great War would have been surprised. I've called it the Lucretius underestimation, after the Latin poetic philosopher who wrote that the fool believes that the tallest mountain there is should be equal to the tallest one he has observed. Danny Kahneman has written, using as backup the works of Howard Kunreuther, that "protective actions, whether by individuals or by governments, are usually designed to be adequate to the worst disaster actually experienced. . . . Images of even worse disaster do not come easily to mind."* For instance, in Pharaonic Egypt, scribes tracked the high-water mark of the Nile and used it as a worst-case scenario. No economist had tested the obvious: Do extreme events fall according to the past? Alas, back-testing says, "No, sorry."

The same dangerous recklessness can be seen in the methodology used for the Fukushima nuclear reactor, built to the worst past outcome and not imagining and extrapolating to much worse. Well, nature, unlike risk engineers, prepares for what has not happened before, assuming worse harm is possible.

So if humans fight the last war, nature fights the next war. Of course, there is a biological limit to our overcompensation.

This form of redundancy remains vastly more extrapolative than our minds, which are intrapolative.

The great Benoit Mandelbrot, now gone for two years, saw the same fractal self-similarity in nature and in probabilities of historical and economic events. It is thrilling to see how the two domains unite under the notion of fractal-based redundancy.

---

* Daniel Kahneman, *Thinking, Fast and Slow* (New York: Farrar, Straus & Giroux, 2011), 137.

P.S.: The word "fitness" in the common scientific discourse does not appear to be precise enough. I am unable to figure out if what is called "Darwinian fitness" is merely intrapolative adaptation to the current environment or if it contains an element of statistical extrapolation. In other words, there is a significant difference between robustness (is not harmed by stressors) and what I've called antifragility (i.e., gains from stressors).

THE BEAUTIFUL LAW OF UNINTENDED CONSEQUENCES

ROBERT KURZBAN

Associate professor of evolutionary psychology, University of Pennsylvania; director, Pennsylvania Laboratory for Experimental Evolutionary Psychology (PLEEP); author, Why Everyone (Else) is a Hypocrite

According to the guide on my walking tour of The Rocks neighborhood in Sydney, Australia, when the plague hit the city around 1900 a bounty was placed on rats to encourage people to kill them, since it was known that rats bore the fleas that communicated the disease to humans. The intent of the bounty was plain enough: reducing the number of rats to reduce the spread of the plague. An unintended consequence, however, was that residents, tempted by the rat bounty, began breeding rats.

The law of unintended consequences is often associated with the American sociologist Robert Merton, though its general spirit appears in various forms, not least in Adam Smith's notion of the Invisible Hand, and is somehow delightful in its chaos, as if Nature were continually thumbing her nose at our attempts to control her.

The idea is that when people intervene in systems with a lot of moving parts—especially ecologies and economies—the intervention, because of the complex interrelationships among the system's parts, will have effects beyond those intended, including many that were unforeseen or unforeseeable.

Examples abound. Returning to Australia, one of the best known examples of an unintended consequence is the case of rabbits, brought by the First Fleet as food, released into the wild for hunting, with the unintended consequence that rabbit populations grew to staggering proportions, causing untold ecological devastation. This in turn led to the development of measures to control the rabbits, including an exceptionally long fence, which had the unintended consequence of guiding three young girls home in the 1930s—which in turn had the unintended consequence of inspiring an award-winning motion picture (*Rabbit-Proof Fence*, 2002).

These chains of consequences occur because making changes to one part of a system with many interacting parts leads to changes in other parts of the system. Because many of the systems we try to influence are complex but incompletely understood—bodies, habitats, markets—there are bound to be consequences that are difficult to predict.

This is not to say that the consequences will always be undesirable. Recently, certain municipalities changed the laws governing the use of marijuana, making it easier to obtain for medical purposes. The law might or might not have reduced the suffering of glaucoma victims, but data from traffic accidents suggest that the change in the legislation did reduce fatalities on the road by about 9 percent. (People substituted marijuana for alcohol and apparently drive better stoned than drunk.) Saving drivers' lives was not the intent of the law, but that was the effect. Another example, smaller in scale though closer to my heart, was the recent abrupt increase of parking rates by a third in University City, Philadelphia, where I work. The intent of the law was to raise revenue to help fund the city's schools. An unintended consequence—because students seem disinclined to pay the higher price—is that I can rely on getting a parking spot when I have to drive to school.

Intervention in any sufficiently complicated system is bound to produce unintended effects. We treat patients with antibiotics, and we select for resistant strains of pathogens. We artificially select for wrinkly-faced bulldogs, and less pleasant traits, such as respiratory problems, come along for the ride. We treat morning sickness with thalidomide, and babies with birth defects follow.

In the economic sphere, most policies have various sorts of knock-on effects, with prohibitions and bans providing some of the most profound examples—including, of course, Prohibition, which itself spun off various consequences, among them, arguably, the rise of organized crime. Because governments typically ban only those things for which people have a taste, when bans do arise people find ways to satisfy those tastes, either through substitutes or black markets, both of which lead to varied consequences. Ban sodas, boost sports-drink sales. Ban the sale of kidneys, spawn an international black market for organs and underground surgeries. Ban the hunting of mountain lions, endanger local joggers.

There is something oddly beautiful about the tendrils of causality in complex systems, holding the same appeal we find in the deliberate inelegance of Rube Goldberg machines. And none of this is to say that the inevitable chances of being surprised by our interventions means that we must give in to pessimism. Rather, it reminds us to exercise caution and humility. As we gradually increase our understanding of large, complicated systems, we will develop new ways to glimpse the unintended consequences of our actions. We already have some guiderails: People will find substitutes for banned or taxed products; removing one species in an ecology typically penalizes populations that prey on them and aids species that compete with them; and so on. So whereas there will probably always be unintended consequences, they needn't be completely unanticipated.

WE ARE WHAT WE DO

TIMOTHY D. WILSON

Sherrell J. Aston Professor of Psychology, University of Virginia; author,
Redirect: The Surprising New Science of Psychological Change

People become what they do. This explanation of how people acquire attitudes and traits dates back to the late British philosopher Gilbert Ryle but was formalized by the social psychologist Daryl Bem in his self-perception theory. People draw inferences about who they are, Bem suggests, by observing their own behavior.

Self-perception theory turns common wisdom on its head. People act the way they do because of their personality traits and attitudes, right? They return a lost wallet because they're honest, recycle their trash because they care about the environment, and pay $5 for a caramel brulée latte because they like expensive coffee drinks. It's evident that behavior emanates from our inner dispositions, but Bem's insight was to suggest that the reverse also holds. If we return a lost wallet, there's an upward tick on our honesty meter. After we drag the recycling bin to the curb, we infer that we really care about the environment. And after purchasing the latte, we assume that we are coffee connoisseurs.

Hundreds of experiments have confirmed the theory and shown when this self-inference process is most likely to operate (for example, when people believe they freely chose to behave as they did and when they weren't sure at the outset how they felt).

Self-perception theory is elegant in its simplicity. But it is also deep, with important implications for the nature of the human mind. Two other powerful ideas follow from it. The first is that we are strangers to ourselves. After all, if we knew our own minds, why would we need to guess what our preferences are from our behavior? If our minds were an open book, we would know exactly how honest we are and how much we like lattes. Instead, we often need to look to our behavior to figure out who we are. Self-perception theory thus anticipated the revolution in psychology in the study of human consciousness, a revolution that revealed the limits of introspection.

But it turns out that we don't just use our behavior to reveal our dispositions—we infer dispositions that weren't there before. Often our behavior is shaped by subtle pressures around us, but we fail to recognize those pressures. Thus, we mistakenly believe that our behavior emanated from some inner disposition. Perhaps we aren't particularly trustworthy and instead returned the wallet in order to impress the people around us. But, failing to realize that, we infer that we're squeaky-clean honest. Maybe we recycle because the city has made it easy to do so (by giving us a bin and picking up every Tuesday) and our spouse and neighbors would disapprove if we didn't. Instead of recognizing those reasons, though, we assume that we should be nominated for the Green Neighbor of the Month Award. Countless studies have shown that people are highly susceptible to social influence but rarely recognize the full extent of that susceptibility, thereby misattributing their compliance to their true desires.

Like all good psychological explanations, self-perception theory has practical uses. It is implicit in several versions of psychotherapy, in which clients are encouraged to change their behavior first, on the assumption that changes in their inner dispositions will follow. It has been used to prevent teenage pregnancies, by getting teens to do community service. The volunteer work triggers a change in their self-image, making them feel more a part of their community and less inclined to engage in risky behaviors. In short, we should all heed Kurt Vonnegut's advice: "We are what we pretend to be, so we must be careful about what we pretend to be."

PERSONALITY DIFFERENCES: THE IMPORTANCE OF CHANCE

SAMUEL BARONDES

Jeanne and Sanford Robertson Professor of Neurobiology and Psychiatry, University of California–San Francisco; author, Making Sense of People

In the golden age of Greek philosophy, Theophrastus, Aristotle's successor, posed a question for which he is still remembered: "Why has it come about that, albeit the whole of Greece lies in the same clime, and all Greeks have a like upbringing, we have not the same constitution of character [personality]?" The question is especially noteworthy because it bears on our sense of who each of us is, and we now know enough to offer an answer: Each personality reflects the activities of brain circuits that gradually develop under the combined direction of the person's unique set of genes and experiences. What makes the implications of this answer so profound is that they lead to the inescapable conclusion that personality differences are greatly influenced by chance events.

Two types of chance events influence the genetic contribution to personality. The first, and most obvious, is the events that brought together the person's mom and dad. Each of them has a particular collection of gene variants—a personal sample of the variants that have accumulated in the collective human genome—and the two parental genetic repertoires set limits on the possible variants that can be transmitted to their offspring. The second chance event is the hit-or-miss union of the particular egg and sperm that make the offspring, each of which contains a random selection of half of the gene variants of each parent. It is the interactions of the resultant unique mixture of maternal and paternal gene variants that play a major part in the twenty-five-year-long developmental process that builds the person's brain and personality. So two accidents of birth—the parents who conceive us, and the egg/sperm combinations that make us—have decisive influences on the kinds of people we become.

But genes don't act alone. Although there are innate programs of gene expression that continue to unfold through early adulthood to direct the construction of rough drafts of brain circuits, these programs are specifically designed to incorporate information from the person's physical and social world. Some of this adaptation to the person's particular circumstances must come at specific developmental periods, called critical periods. For example, the brain circuits that control the characteristic intonations of a person's native language are open for environmental input only during a limited window of development.

And just as chance influences the particular set of genes we are born with, so does it influence the particular environment we are born into. Just as our genes incline us to be more or less friendly, or confident, or reliable, the worlds we grow up in incline us to adopt particular goals, opportunities, and rules of conduct. The most obvious aspects of these worlds are cultural, religious, social, and economic, each transmitted by critical agents: parents, siblings, teachers, and peers. And the specific content of these important influences—the specific era, place, culture, etc., that we happen to have been born into—is as much a toss of the dice as the specific content of the egg and sperm that formed us.

Of course, chance is not fate. Recognizing that chance events contribute to individual personality differences doesn't mean that each life is predetermined or that there is no free will. The personality arising through biological and sociocultural accidents of birth can be deliberately modified in many ways, even in maturity.

Nevertheless, the chance events that direct brain development in our first few decades leave enduring residues.

When thinking about a particular personality, it is therefore helpful to be aware of the powerful role chance played in its construction. Recognizing the importance of chance in our individual differences doesn't just remove some of their mystery—it can also have moral consequences by promoting understanding and compassion for the wide range of people with whom we share our lives.

METABOLIC SYNDROME: CELL ENERGY ADAPTATIONS IN A TOXIC WORLD?

BEATRICE GOLOMB

Professor of medicine, University of California–San Diego

Metabolic syndrome (MetSyn) has been called the epidemic of the 21st century. MetSyn is an accretion of symptoms, including high body-mass index, high blood sugar, high blood pressure, high blood triglycerides, high waist circumference, and/or reduced HDL cholesterol, the so-called good cholesterol. Epidemics of obesity and diabetes are intertwined with and accompany the meteoric rise in MetSyn.

The prevalent view is that MetSyn is due to a glut of food calories ("energy") consumed and a dearth of exercise energy expended, spurring weight gain—an "energy surfeit"—with the other features arising in consequence. After all, we have more access to calories, and are more often sedentary, than in times gone by. MetSyn factors are each linked, in otherwise healthy young populations, to higher mortality.

But this view leaves many questions unanswered: Why do elements of MetSyn correlate? Why are overweight people today more likely to have diabetes than hitherto? Why are elements of MetSyn now emerging in infancy? Why is MetSyn materializing in poor and third-world nations?

The customary explanation also creates paradoxes. If MetSyn stems from energy surfeit, why do the following factors, which reduce energy supply or increase demand, promote MetSyn—far from protecting against it?

- Sleep apnea (It's a stronger MetSyn risk factor than being overweight; moreover, sleep-apnea treatment benefits each MetSyn element)
- Ultra low-calorie or low-fat diets
- Fasting, skipped meals
- Hypoglycemia-promoting diets (high-carbohydrate/low-fat/low-protein diets lead to unopposed insulin surge)
- Deficient sleep (more energy-expending wake time)
- Illness/injury/surgery (high energy demand)
- Cold weather (mandating energy expenditure for thermogenesis)
- Nutrient and antioxidant deficiencies (adequacy required for energy production)
- Oxidative stressor exposure (impairs function of mitochondria, the energy-producing elements of cells)
- Mitochondrial pathology

Why do factors that protect from energy-deficit (antioxidant cocoa and cinnamon, mitochondria-supportive coenzyme Q10) reduce MetSyn factors?

Why does exercise, which expends energy but boosts energy production via antioxidant effects, mitochondrial biogenesis, enhanced circulation, and cardiopulmonary function (improving oxygen intake, delivery, and conversion to energy) reduce MetSyn factors?

Why does MetSyn cease to elevate mortality (indeed, sometimes boost survival) when the group studied is of advanced age or suffers from heart failure or severe kidney disease—all conditions that impair cell energy?

Suppose the correct explanation were the complete opposite of the accepted one. Could the features of MetSyn be the adaptive response to *inadequate* energy? After all, fat depots, glucose, and triglycerides are each accessory energy sources (oxygen is primary); blood pressure is needed to deliver these to tissues, especially when underperfused. Cell energy, central to cell and organism survival, is needed continuously; we live only minutes without oxygen. The stretch is not so great: Populations in which prior generations were energy-starved now have increased obesity/MetSyn, and low energy supply *in utero* is understood to foster MetSyn in adulthood.

This explains, as the energy-surfeit view does not, why MetSyn exists at all: why elevated glucose, triglycerides, blood pressure (carrying oxygen, glucose, nutrients), and abdominal fat deposition cohere statistically. It explains why other energy-supportive adaptations, like free fatty acids and (metabolically active) ectopic fat (fatty-liver, fatty-pancreas, fatty-kidney, even fatty-streaks in the blood vessels) accompany those Met Syn factors; why MetSyn is linked to fatigue and increased sleep duration (which conserves energy). Indeed, increased caloric intake and reduced exercise— the usual MetSyn explanation—arise, too, as fellow energetic adaptations. Thus, this view is arguably not antithetical to the canonical one but in one sense subsumes it. It explains, as the energy-surfeit view does not, the populations at risk for MetSyn, such as the elderly (mitochondrial function declines exponentially with age) and those afflicted with sleep apnea or any cause for recurrent impairment of energy production. And it explains why, in studies focused on persons with conditions that blight energy, those with MetSyn features paradoxically don't do worse, or even fare better.

Why, then, is MetSyn an epidemic? Numerous energy-adverse secular shifts include nutrient-poor/low-antioxidant/high-prooxidant pseudofoods (nutrients are needed for energy-production machinery, and antioxidants protect from oxidative stress, whose chief target is mitochondria, the energy-producing bits of cells), decline of regular balanced meals, and hypoglycemic-promoting macronutrient composition (simple carbohydrates without fat or protein engender unopposed insulin surges, and glucose drops). But a central factor is the explosion in our environment of oxidative stressors that disrupt function of (and DNA in) mitochondria. For example:

- Metals and heavy metals (mercury in fish, high-fructose corn syrup, broken lightbulbs; arsenic to promote poultry growth; aluminum vaccine adjuvants with proliferating childhood vaccinations)
- Plastics with bisphenol A
- Personal-care products (chemicals in sunscreens, lotions, hair dyes, cosmetics, detergents, fabric softeners and dryer sheets, conditioners)
- Cleaning products
- Furnishings and clothes with formaldehyde (pressboard, no-iron cotton)
- Petrochemicals, combustion products
- Electromagnetic fields (electronics, cell phones, smart-meters)
- Fire retardants (pajamas, bedding)
- Dry-cleaning chemicals
- Air "fresheners"
- Pesticides, herbicides (potent oxidative stressors, now routinely applied in homes and office buildings and at recreational sites)
- Termite tenting
- Prescription and over-the-counter drugs, including antibiotics—both direct exposure and through our food supply
- Antimicrobial soaps with active ingredients largely unfilterable from the water supply

- Air and water pollutants and contaminants
- Artificial ingredients in foods—transfats, artificial sweeteners, dyes, preservatives

The energy-deficit ("starving cell") hypothesis accounts for scores of facts for which the prevailing view provides no insight. Observations deemed paradoxical in the standard view emerge seamlessly. The hypothesis makes testable predictions—for example, that many other oxidative stress- and mitochondrial-disruption-inducing exposures that have not yet been assessed will promote one or more elements of MetSyn. For factors that relate at both extremes to MetSyn—say, short and long sleep duration— the energy-disruptive one will prove to cause MetSyn and the energy-supportive one to serve as a fellow adaptive consequence.

This reframing addresses an important problem. Some think MetSyn is slated to reverse the gains we have made in longevity. The conclusion might be surprising—and should precipitate a revision in our thinking not just about the causes of MetSyn but also about its solutions.

DEATH IS THE FINAL REPAYMENT

EMANUEL DERMAN

Professor of professional practice, Columbia University; former managing director, Goldman Sachs; author, Models. Behaving. Badly.

"Sleep is the interest we have to pay on the capital which is called in at death; and the higher the rate of interest and the more regularly it is paid, the further the date of redemption is postponed."

So wrote Arthur Schopenhauer, comparing life to finance in a universe that must keep its books balanced. At birth you receive a loan—consciousness and light borrowed from the void, leaving a hole in the emptiness. The hole will grow bigger each day. Nightly, by yielding temporarily to the darkness of sleep, you restore some of the emptiness and keep the hole from growing limitlessly. In the end, you must pay back the principal, complete the void, and return the life originally lent you.

By focusing on the common periodic nature of sleep and interest payments, Schopenhauer extends the metaphor of borrowing to life itself. Life and consciousness are the principal, death is the final repayment, and sleep is *la petite mort*, the periodic little death that renews.

DENUMERABLE INFINITIES AND MENTAL STATES

DAVID GELERNTER

Computer scientist, Yale University; chief scientist, Mirror Worlds Technologies; author, America-Lite: How Imperial Academia Dismantled Our Culture (and Ushered In the Obamacrats)

My favorites:

1. The 19th-century German mathematician Georg Cantor's explanation of why all denumerable infinities are the same size—why, for example, the set of all integers is the same size as the set of all positive integers, or all even integers—and why some infinities are bigger than others. (The set of all rational numbers is the same size as the set of all integers, but the set of all real numbers—terminating plus nonterminating decimals—is larger.) The set of all positive integers is the same size as the set of all even, positive integers—to see that, just line them up, one by one. 1 is paired with 2 (the first even positive integer), 2 is paired with 4, 3 with 6, 4 with 8, and so on. You'd think there would be more positive integers than even ones; but this pairing-off shows that no positive integer will ever be left without a partner. (And so they all dance happily off and there are no wallflowers.) The other proofs are similar in their stunning simplicity, but much easier to demonstrate on a blackboard than describe in words.

2. Equally favorite: Philosopher John Searle's proof that no digital computer can have mental states (a mental state is,

example, your state of mind when I say, "Picture a red rose" and you do)—that minds can't be built out of software. A digital computer can do only trivial arithmetic and logical instructions. You can do them, too; you can execute any instruction that a computer can execute. You can also imagine yourself executing lots and lots of trivial instructions. Then ask yourself, "Can I picture a new mind emerging on the basis of my doing lots and lots and lots of trivial instructions?" No. Or imagine yourself sorting a deck of cards—sorting is the kind of thing digital computers do. Now imagine sorting a bigger and bigger and bigger deck. Can you see consciousness emerging at some point, when you sort a large enough batch? Nope.

And the inevitable answer to the inevitable first objection: But neurons only do simple signal transmission—can you imagine consciousness emerging out of *that*? This is an irrelevant question. The fact that lots of neurons make a mind has no bearing on the question of whether lots of anything else make a mind. I can't imagine being a neuron, but I can imagine executing machine instructions. No mind emerges, no matter how many of those instructions I carry out.

INVERSE POWER LAWS

RUDY RUCKER
*Mathematician, computer scientist; cyberpunk
pioneer; novelist; author,* Surfing the Gnarl

I'm intrigued by the empirical fact that most aspects of our world and our society are distributed according to so-called inverse power laws. That is, many distribution curves take on the form of a curve that swoops down from a central peak to have a long tail that asymptotically hugs the horizontal axis.

Inverse power laws are elegantly simple and deeply mysterious, but more galling than beautiful. Inverse power laws are self-organizing and self-maintaining. For reasons that aren't entirely understood, they emerge spontaneously in a wide range of parallel computations, both social and natural.

One of the first social scientists to notice an inverse power law was the philologist George Kingsley Zipf, who formulated an observation now known as Zipf's Law. This is the statistical fact that in most documents the frequency with which a given word is used is roughly proportional to the reciprocal of the word's popularity rank. Thus, the second most popular word is used half as much as the most popular word, the tenth most popular word is used a tenth as much as the most popular word, and so on.

In society, similar kinds of inverse power laws govern society's rewards. Speaking as an author, I've noticed, for instance, that the hundredth most popular author sells a hundredfold fewer books than the author at the top. If the top writer sells a million copies, somone like me might sell 10,000.

Disgruntled scribes sometimes fantasize about a utopian marketplace in which the naturally arising inverse-power-law distri-

ould be forcibly replaced by a linear distribution—that
ales schedule that lies along a smoothly sloping line instead
of taking the form of the present bent curve that starts at an
impudently high peak and swoops down to dawdle along the horizontal axis.

But there's no obvious way that the authors' sales curve could
be changed. Certainly there's no hope of having some governing
group try to force a different distribution. After all, people make
their own choices as to what books to read. Society is a parallel
computation, and some aspects of it are beyond control.

The inverse-power-law aspects of income distribution are particularly disturbing. Thus the second wealthiest person in a society might own half as much as the richest, with the tenth richest
person possessing only a tenth as much, and—out in the 'burbs—
the thousandth richest person is making only one-thousandth as
much as the person on the top.

Putting the same phenomenon a little more starkly, while a
company's chief executive officer might earn $100,000,000 a year,
a software engineer at the same company might earn only $100,000
a year, and a worker in one of the company's overseas assembly
plants might earn only $10,000 a year—a ten-thousandth as much
as the top exec.

Power-law distributions can also be found in the opening week-
end grosses of movies, in the number of hits that Web pages get,
and in the audience shares for TV shows. Is there some reason the
top ranks do so overly well and the bottom ranks seem so unfairly
penalized? The short answer is no, there's no real reason. There
need be no conspiracy to skew the rewards. Galling as it seems,
inverse-power-law distributions are a fundamental natural law
about the behavior of systems. They're ubiquitous.

Inverse power laws aren't limited to societies—they also dominate the statistics of the natural world. The tenth smallest lake is

likely to be a tenth as large as the biggest one, the hundredth larg est tree in a forest may be a hundredth as big as the largest tree, the thousandth largest stone on a beach is a thousandth the size of the largest one.

Whether or not we like them, inverse power laws are as inevitable as turbulence, entropy, or the law of gravity. This said, we can somewhat moderate them in our social context, and it would be too despairing to say we have no control whatsoever over the disparities between our rich and our poor.

But the basic structures of inverse-power-law curves will never go away. We can rail at an inverse power law if we like—or we can accept it, perhaps hoping to bend the harsh law toward not so steep a swoop.

HE LEOPARD GOT HIS SPOTS

SAMUEL ARBESMAN

Applied mathematician; senior scholar, Ewing
Marion Kauffman Foundation

In one of his celebrated just-so stories, Rudyard Kipling recounted how the leopard got his spots. Taking this approach to its logical conclusion, we would need distinct stories for every animal's pattern: the leopard's spots, the cow's splotches, the panther's solid color. And we would have to add even more stories for the complex patterning of everything from molluscs to tropical fish.

But far from these different animals requiring separate and distinct explanations, there is a single underlying explanation that shows how we can get all of these different patterns using a single unified theory.

Beginning in 1952, with Alan Turing's publication of a paper entitled "The Chemical Basis of Morphogenesis," scientists recognized that a simple set of mathematical formulas could dictate the variety of ways that patterns and colorings form in animals. This model is known as a reaction-diffusion model and works in a simple way: Imagine that you have multiple chemicals, which diffuse over a surface at different rates and can interact. Whereas in most cases diffusion simply creates a uniformity of a given chemical—think how cream poured into coffee will eventually spread and dissolve and create a lighter brown liquid—when multiple chemicals diffuse and interact it can give rise to nonuniformity. Although this is somewhat counterintuitive, it not only occurs but also can be generated using only a simple set of equations—thus explaining the exquisite variety of patterns seen in the animal world.

Mathematical biologists have been exploring the properties of reaction-diffusion equations ever since Turing's paper. They've found that varying the parameters can generate the animal patterns we see. Some mathematicians have examined the ways in which the size and shape of the surface can dictate the patterns we see; as the size parameter is modified, we can easily go from such patterns as giraffe-like to those seen on Holstein cows.

This elegant model can even yield simple predictions. For example, whereas a spotted animal can have a striped tail (and very often does) according to the model, a striped animal will never have a spotted tail. And this is exactly what we see! These equations can generate the endless variation seen in nature but also show the limitations inherent in biology. The just-so of Kipling may be safely exchanged for the elegance and generality of reaction-diffusion equations.

THE UNIVERSAL ALGORITHM FOR HUMAN DECISION MAKING

STANISLAS DEHAENE

Neuroscientist, Collège de France; author, Reading in
the Brain: The New Science of How We Read

The ultimate goal of science, as the French physicist Jean Baptiste Perrin once stated, should be "to substitute visible complexity for an invisible simplicity." Can human psychology achieve this ambitious goal: the discovery of elegant rules behind the apparent variability of human thought? Many scientists still consider psychology a "soft" science, whose methods and object of study are too fuzzy, too complex, and too suffused with layers of cultural complexity to ever yield elegant mathematical generalizations. Yet cognitive scientists know that this prejudice is wrong. Human behavior obeys rigorous laws of the utmost mathematical beauty and even necessity. I will nominate just one of them: the mathematical law by which we make our decisions.

All of our mental decisions appear to be captured by a simple rule that weaves together some of the most elegant mathematics of the past centuries: Brownian motion, Bayes' Law, and the Turing machine. Let's start with the simplest of all decisions: How do we decide that 4 is smaller than 5? Psychological investigation reveals many surprises behind this simple feat. First, our performance is slow: The decision takes us nearly half a second, from the moment the digit 4 appears on a screen to the point when we respond by clicking a button. Second, our response time is highly variable from trial to trial, anywhere from 300 milliseconds to 800 milliseconds, even though we are responding to the very same digital symbol, "4." Third, we make errors—it sounds ridiculous, but even when comparing 4 with 5 we sometimes make the wrong decision. Fourth, our performance varies with the meaning of the objects: We are much faster and make fewer errors when the numbers are far from each other (such as 1 and 5) than when they are close (such as 4 and 5).

All of the above facts, and many more, can be explained by a single law: Our brain makes decisions by accumulating the available statistical evidence and committing to a decision whenever the total exceeds a threshold.

Let me unpack this statement. The problem the brain faces when making a decision is one of sifting the signal from the noise. The input to any of our decisions is always noisy: Photons hit our retina at random times, neurons transmit the information with partial reliability, and spontaneous neural discharges (spikes) are emitted throughout the brain, adding noise to any decision. Even when the input is a digit, neuronal recordings show that the corresponding quantity is coding by a noisy population of neurons that fires at semi-random times, with some neurons signaling "I think it's 4," others "it's close to 5" or "it's close to 3," and so on. Because the brain's decision system sees only unlabeled spikes, not full-fledged symbols, separating the wheat from the chaff is a genuine problem for it.

In the presence of noise, how should one make a reliable decision? The mathematical solution to that problem was first addressed by Alan Turing, when he was cracking the Enigma code at Bletchley Park. Turing found a small glitch in the Enigma machine, which meant that some of the German messages contained small amounts of information, but unfortunately too little for him to be certain of the underlying code. He realized that Bayes' Law could be exploited to combine all the independent pieces of evidence. Skipping the math, Bayes' Law provides a simple way to sum all of the successive hints, plus whatever prior knowledge we have, in order to obtain a combined statistic that tells us what the total evidence is.

With noisy inputs, this sum fluctuates up and down, as some incoming messages support the conclusion while others merely add noise. The outcome is what mathematicians call a random walk—a fluctuating march of numbers as a function of time. In our case, however, the numbers have a currency: They represent the likelihood that one hypothesis is true (e.g., the probability that the input digit is smaller than 5). Thus, the rational thing to do is to act as a statistician and wait until the accumulated statistic exceeds a threshold probability value. Setting it to $p = 0.999$ would mean that we have 1 chance in 1,000 to be wrong.

Note that we can set this threshold to any arbitrary value. However, the higher we put it, the longer we have to wait for a decision. There is a speed/accuracy tradeoff: We can wait a long time and make a highly accurate but conservative decision, or we can hazard a response earlier but at the cost of making more errors. Whatever our choice, we will always make a few errors.

Suffice it to say that the decision algorithm I sketched—which simply describes what any rational creature should do in the face of noise—is now considered a fully general mechanism for human decision making. It explains our response times, their variability, and the entire shape of their distribution. It describes why we make errors, how errors relate to response time, and how we set the speed/accuracy tradeoff. It applies to all sorts of decisions, from sensory choices (Did I see movement or not?) to linguistics (Did I hear "dog" or "bog"?) to higher-level conundrums (Should I do this task first or second?). And in more complex cases, such as performing a multidigit calculation or a series of tasks, the model characterizes our behavior as a sequence of accumulate-and-threshold steps, which turns out to be an excellent description of our serial, effortful Turing-like computations.

Furthermore, this behavioral description of decision making is now leading to major progress in neuroscience. In the monkey brain, neurons can be recorded whose firing rates index an accumulation of relevant sensory signals. The theoretical distinction between evidence accumulation and threshold helps parse out the brain into specialized subsystems that make sense from a decision-theoretic viewpoint.

As with any elegant scientific law, many complexities are waiting to be discovered. There is probably not just one accumulator but many, as the brain accumulates evidence at each of several successive levels of processing. Indeed, the human brain increasingly fits the bill for a superb Bayesian machine that makes massively parallel inferences and microdecisions at every stage. Many of us think our sense of confidence, stability, and even conscious awareness may result from such higher-order cerebral "decisions" and will ultimately fall prey to the same mathematical model. Valuation is also a key ingredient, one that I skipped, although it demonstrably plays a crucial role in weighing our decisions. Finally, the system is ripe with a-prioris, biases, time pressures, and other top evaluations that draw it away from strict mathematical optimality.

Nevertheless, as a first approximation, this law stands as one of the most elegant and productive discoveries of 20th-century psychology: Humans act as near-optimal statisticians, and our decisions correspond to an accumulation of the available evidence up to some threshold.

LORD ACTON'S DICTUM

MIHALY CSIKSZENTMIHALYI

Distinguished Professor of Psychology and Management, Claremont Graduate University; founding codirector of CGU's Quality of Life Research Center; author, Flow: The Psychology of Optimal Experience

I hope I will not be drummed out of the corps of social science if I confess that I can't think of an explanation in our field that is both elegant and beautiful. Perhaps deep. . . . I guess we are still too young to have explanations of that sort. But there is one elegant and deep statement (which, alas, is not quite an "explanation") that comes close to fulfilling the *Edge* Question criteria and that I find very useful as well as beautifully simple.

I refer to the well-known lines Lord Acton wrote in a letter from Naples in 1887 to the effect that "Power tends to corrupt, and absolute power corrupts absolutely." At least one philosopher of science has written that on this sentence an entire science of human beings could be built.

I find that the sentence offers the basis for explaining how a failed painter like Adolf Hitler and a failed seminarian like Joseph Stalin could end up with the blood of millions on their hands; or how the Chinese emperors, the Roman popes, and the French aristocracy failed to resist the allure of power. When a religion or ideology becomes dominant, the lack of controls will result in widening spirals of license, leading to degradation and corruption.

It would be nice if Acton's insight could be developed into a full-fledged explanation before the hegemonies of our time, based on blind faith in science and the worship of the Invisible Hand, follow older forms of power into the dustbins of history.

FACT, FICTION, AND OUR PROBABILISTIC WORLD

VICTORIA STODDEN

Computational legal scholar; assistant professor of statistics, Columbia University

How do we separate fact from fiction? We are frequently struck by seemingly unusual coincidences. Imagine seeing an inscription describing a fish in your morning reading, and then at lunch you are served fish and the conversation turns to "April fish" (or April fools). That afternoon, a work associate shows you several pictures of fish, and in the evening you are presented with an embroidery of fishlike sea monsters. The next morning, a colleague tells you she dreamed of fish. This might start to seem spooky, but it turns out that we shouldn't find it surprising. The reason has a long history, resulting in the unintuitive insight of building randomness directly into our understanding of nature, through the probability distribution.

Chance as Ignorance

Tolstoy was skeptical of our understanding of chance. He gave an example of a flock of sheep, one of which had been chosen for slaughter. This one sheep was given extra food separately from the others, and Tolstoy imagined that the flock, with no knowledge of what was coming, must find the continually fattening sheep extraordinary—something he thought they would assign to chance due to their limited viewpoint. Tolstoy's solution was for the flock of sheep to stop thinking that things happen only for "the attainment of their sheep aims" and realize that there are hidden aims that explain everything perfectly well, and so no need to resort to the concept of chance.

Chance as an Unseen Force

Eighty-three years later, Carl Jung published a similar idea in his well-known essay "Synchronicity, An Acausal Connecting Principle." He postulated the existence of a hidden force responsible for the occurrence of seemingly related events that otherwise appear to have no causal connection. The initial story of the six fish encounters is Jung's, taken from his book. He finds this string of events unusual—too unusual to be ascribable to chance. He thinks something else must be going on and labels it the acausal connecting principle.

Persi Diaconis, Mary V. Sunseri Professor of Statistics and Mathematics at Stanford and a former professor of mine, thinks critically about Jung's example: Suppose we encounter the concept of fish once a day on average, according to what statisticians call a Poisson process (another fish reference!). The Poisson process is a standard mathematical model for counts—for example, radioactive decay seems to follow a Poisson process. The model presumes a certain fixed rate at which observations appear on average, and otherwise they are random. So we can consider a Poisson process for Jung's example with a long-run average rate of one observation per twenty-four hours and calculate the probability of seeing six or more observations of fish in a twenty-four-hour window. Diaconis finds the chance to be about 22 percent. Seen from this perspective, Jung shouldn't have been surprised.

The Statistical Revolution: Chance in Models of Data Generation

Only about two decades after Tolstoy penned his lines about sheep, the English mathematician Karl Pearson brought about a statistical revolution in scientific thinking with a new idea of how observations arose—the same idea used by Diaconis in his probability calculation. Pearson suggested that nature presents data from an unknown distribution but with some random scatter. His insight was that this is a different concept from measurement error, which adds additional error when the observations are actually recorded.

Before Pearson, science dealt with things that were "real," such as laws describing the movement of the planets or blood flow in horses (to use examples from David Salsburg's book, *The Lady Tasting Tea*). What Pearson made possible was a probabilistic conception of the world. Planets didn't follow laws with exact precision, even after accounting for measurement error. The exact course of blood flow differed in different horses, but the horse circulatory system wasn't purely random. In estimating distributions rather than the phenomena themselves, we are able to abstract a more accurate picture of the world.

Chance Described by Probability Distributions

That measurements themselves have a probability distribution was a marked shift from confining randomness to the errors in the measurement. Pearson's conceptualization is useful because it permits us to estimate whether what we see is likely or not, under the assumptions of the distribution. This reasoning is now our principal tool for judging whether or not we think an explanation is likely to be true.

We can, for example, quantify the likelihood of drug effectiveness or carry out particle detection in high-energy physics. Is the distribution of the mean-response difference between drug treatment and control groups centered at zero? If that seems likely, we can be skeptical of the drug's effectiveness. Are candidate signals so far from the distribution for known particles that they must be from a different distribution, suggesting a new particle? Detecting the Higgs boson requires such a probabilistic understanding of the data, to differentiate Higgs signals from other events. In all these cases, the key is that we want to know the characteristics of the underlying distribution that generated the phenomenon of interest.

Pearson's incorporation of randomness directly into the probability distribution enables us to think critically about likelihoods and quantify our confidence in particular explanations. We can better evaluate when what we see has special meaning and when it does not, permitting us to better reach our "human aims."

ELEGANT = COMPLEX

GEORGE CHURCH
Professor of genetics, Harvard Medical School;
director, Personal Genome Project

Many would say the opposite, elegance = simplicity. They have (classical) physics envy—for smooth, linear physics and describably delicious four-letter words, like $F=ma$. But modern science has moved on, embracing the complex. Occam now uses a Web-enabled fractal e-razor. Even in mathematics, stripped of the awkward realities of nonideal gases, turbulence, and nonspherical cows, simple statements about integers like Fermat's $a^n + b^n = c^n$ and wrangling maps with four colors take many years and pages (occasionally computers) to prove.

The question is not "What is your favorite elegant explanation?" but "What should your favorite elegant explanation be?" We're capable of changing not only our minds but also the whole fabric of human nature. As we engineer, we recurse—successively approximating ourselves as an increasingly survivable force of nature. If so, what will we ultimately admire? Our evolutionary baggage served our ancestors well but could kill our descendants. Faced with modern foods, our frugal metabolisms feed a diabetes epidemic. Our love of "greedy algorithms" leads to exhausted resources. Our too-easy switching from rationality to blind faith or fear-based decision making can be manipulated politically to drive conspicuous consumption. (Think Easter Island, 163 square kilometers of devastation, scaled to Earth Island at 510 million square kilometers.) "Humans" someday may be born with bug-fixes for dozens of current cognitive biases, as well as intuitive understanding and motivation to manipulate quantum weirdnesses, dimensions beyond three, super rare events, global economics, etc. Agricultural and cultural monocultures are evolutionarily bankrupt. Evolution was only briefly focused on surviving in a sterile world of harsh physics, but ever since has focused on life competing with itself. Elegant explanations are those that predict the future farther and better. Our explanations will help us dodge asteroids, solar red-giant flares, and even close encounters with the Andromeda galaxy. But most of all, we will diversify to deal with our own ever-increasing complexity.

TINBERGEN'S QUESTIONS

IRENE PEPPERBERG

Research associate & lecturer, Harvard University; adjunct associate professor of psychology, Brandeis University; author, Alex & Me

Why do we—and all other creatures—behave as we do? No answers really exist. I chose the ethologist and ornithologist Nikolaas Tinbergen's questions for exactly that reason, because sometimes there is no one deep, elegant, and beautiful explanation. Much like a teacher of fishing rather than a giver of fish, Tinbergen did not try to provide a global explanation but instead gave us a scaffolding upon which to build our own answers to each individual behavioral pattern we observe—a scaffolding that can be used not only for the ethological paradigms for which he was famous but also for all forms of behavior in any domain. Succinctly, Tinbergen asked:

- What is the mechanism? How does it seem to work?
- What is the ontogeny? How do we observe it develop over time?
- What is its function? What are all the possible reasons it is done?
- What is its origin? What are the many ways in which it could have arisen?

In attempting to answer each of these questions, we are forced to think, at the very least, about the interplay of genes and environment, of underlying processes (neuroanatomy, neurophysiology, hormones, and so on), of triggers and timing, what advantages and disadvantages are balanced, and how these may have changed over time.

Furthermore, unlike most "favorite" explanations, Tinbergen's questions are enduring. Answers to his questions often reflect a current zeitgeist in the scientific community, but those answers mutate as additional knowledge becomes available. His questions challenge us to rethink our basic presumptions each time another chunk of data lands in our laps, whatever our field of study. Our fascination with simple elegant answers strikes me as a Douglas Adams (*Hitchhiker's Guide to the Galaxy*) pursuit: We may find "42," but unless we know how to formulate the appropriate questions, the answer isn't always very meaningful.

THE UNIVERSAL TURING MACHINE

GLORIA ORIGGI

Philosopher, Centre National de la Recherche Scientifique, Paris; editor, Text-e: Text in the Age of the Internet

"There are more things in heaven and earth, Horatio, than are dreamt of in your philosophy," says Hamlet to his friend Horatio. An elegant way to point to all the unsolvable, untreatable questions that haunt our lives. One of the most wonderful demonstrations of all time ends up with the same sad conclusion: Some mathematical problems are simply unsolvable.

In 1936, the British mathematician Alan Turing conceived the simplest and most elegant computer ever, a device (as he later described it in a 1948 essay titled "Intelligent Machinery") with

> an infinite memory capacity obtained in the form of an infinite tape marked out into squares, on each of which a symbol could be printed. At any moment there is one symbol in the machine; it is called the scanned symbol. The machine can alter the scanned symbol and its behaviour is in part determined by that symbol, but the symbols on the tape elsewhere do not affect the behaviour of the machine. However the tape can be moved back and forth through the machine, this being one of the elementary operations of the machine.

An abstract machine, conceived by the mind of a genius, to solve an unsolvable problem: the *decision problem.* That is, for each logical formula in a theory, is it possible to decide in a finite number of steps if the formula is valid in that theory? Well, Turing shows that it's not possible. The decision problem, or *Entscheidungsproblem*, was well known by mathematicians: It was the tenth on a list of unsolved problems that David Hilbert presented in 1900 to the mathematics community, thus setting most of the 20th century's agenda for mathematical research. It asks whether there is a *mechanical* process realizable in a finite number of steps that can decide whether a formula is valid or not or whether a function is computable or not. Turing began by asking himself, "What does a *mechanical process* mean?" and his answer was that a mechanical process is a process that can be realized by a machine. Obvious, isn't it?

He then designed a machine for each possible formula in first-order logic and for each possible recursive function of natural numbers—given the logical equivalence proved by Gödel, in his incompleteness theorem, between the set of first-order-logic formulas and the set of natural numbers. And, indeed, using Turing's simple definition, we can write down a string of 0s and 1s for each tape to describe a function, then give to the machine a list of simple instructions (move left, move right, stop) so that it writes down the demonstration of the function and then stops.

This is his Universal Turing Machine—universal because it can take as input any possible string of symbols describing a function and give as output its demonstration. But if you feed the Universal Turing Machine with a description of itself, it doesn't stop; it goes on infinitely generating 0s and 1s. That's it. The Mother of all computers, the soul of the digital age, was designed to show that not everything can be reduced to a Turing machine. There are more things in heaven and earth than are dreamt of in our philosophy.

A MATTER OF POETICS

RICHARD FOREMAN
Playwright and director; founder, Ontological-Hysteric Theater

Since every explanation is contingent, limited by its circumstances, and certain to be superseded by a better or momentarily more ravishing one, the favorite explanation is really a matter of poetics rather than science or philosophy. That being said, I, like everyone else, fall in "love"—a romantic infatuation that either passes or transforms into something else. But it is the *repeated* momentary ravishment that slowly shapes one, because, in a sense, one is usually falling in love with the same type again and again, and this repetition defines and shapes one's mental character. When young, I was so shaped and oriented by what I shall now call my *two* favorite explanations.

1. I hardly remember the details (being no scientist, of course), but I remember reading about Paul Dirac's theory of the sea of negative energy out of which popped—by a hole, an absence—the positron, which built the world we know. I hope I have this right and haven't made a fool of myself by misrepresenting it—but, in a sense, that wouldn't matter. Because this image, fueled by this explanation, energized my exploration into a new kind of theater in which (evoking a kind of negative theology) I tried, and still try, to pull an audience into the void rather than feeding it what it already feels about the "real" world and wants confirmed.

2. Shortly thereafter (this was all in the 1950s), encountering the unjustly neglected philosopher Ortega y Gasset and

being sent spinning by his explanation that a human being is not a "whole persona" (in a world where the mantra was to become "well rounded") but, as he famously put it, "I am myself and my circumstances"—that is, a split creature.

And what Ortegean circumstantial setting set me up to be seduced by the Dirac-ian explanation? It had something to do with growing up in privileged Scarsdale, hating it but hiding the fact that I felt out of place and awkward by becoming a successful high-school achiever. Dirac's (for me) powerful poetic metaphor let me imagine that the unreachable source (the sea of negative energy) was the real ground on which we all secretly stood—and to take courage from the fact that the world surrounding me was blind to the deeper reality of things and that my alienation was in some sense justified.

THE ORIGINS OF BIOLOGICAL ELECTRICITY

JARED DIAMOND

*Professor of geography, University of California–Los Angeles;
author,* Collapse: How Societies Choose to Fail or Succeed

My favorite deep, elegant, and beautiful explanation is the solution to the problem of the biological generation of electricity by animals and plants, provided by the British physiologists Alan Hodgkin and Andrew Huxley in 1952, for which they received the Nobel Prize in physiology or medicine in 1963.

It had been known for over a century that nerves, muscles, and some other organs of animals and a few plants generate electricity. Most of that electricity is at low voltages of just a fraction of a volt. However, electric eels arrange 6,000 muscle membranes in series and thereby generate 600 volts, enough to kill their prey, shock horses wading rivers, and shock me when I was studying eel electricity generation as a graduate student and got so focused on thinking about physiological mechanisms that I forgot their consequences.

Electricity involves the movement of charged particles. In our light bulbs and electric grids, those charged particles are negatively charged electrons. What are they in biological systems? Already over a century ago, the German physiologist Julius Bernstein speculated that the charged particles whose motion was responsible for biological electricity were not electrons but positively charged ions.

Hodgkin and Huxley started the decisive experiments in the late 1930s. They expected to find that the voltage across a resting nerve membrane went transiently to zero during an electrical impulse, due to a loss of selective permeability to the positively charged potassium ion. To their surprise, they found that the nerve voltage didn't just go to zero, and the nerve membrane didn't just become indiscriminately permeable: The voltage actually reversed in sign, requiring something special. But then Hitler invaded Poland, and Hodgkin and Huxley spent the next six years using their understanding of electricity to build radar sets for the British military.

In 1945, they resumed their experiments, using giant nerves that had been discovered in the backs of squid and that were big enough to insert an electrode in for measuring the voltage across the nerve membrane. They confirmed their tantalizing prewar discovery that the nerve voltage really did reverse in sign and that that reversal got transmitted along a nerve to constitute an electrical impulse. In a series of experiments that define the word "elegance," they then artificially clamped the voltage across the nerve membrane at various levels, measured the electric currents going in and out of the membrane as a function of time at each level after the voltage clamp, translated those voltage measurements into changes of permeability to the positively charged potassium ion and then to the positively charged sodium ion as a function of voltage and time, and finally reconstructed the whole course of a nerve impulse from those time-dependent and voltage-dependent permeability changes. Today, physiology students do the necessary calculations to reconstruct an action potential in an afternoon on their desk computers. In 1952, before the era of modern computers, Andrew Huxley had to do the calculations much more laboriously, with a desk calculator: It took him about a month to calculate one nerve impulse.

The four papers that Hodgkin and Huxley published in the British *Journal of Physiology* in 1952 were so overwhelming in their detailed unraveling of sodium and potassium movements, and their reconstruction of nerve impulses, that the scientific world became convinced almost immediately. Those permeability changes to positive ions (not to negative electrons) make it possible for nerves to convey electrical impulses, for muscles to convey impulses that activate contraction, for nerve/muscle junctions to convey impulses by which nerves activate muscles, for nerve/nerve junctions (so-called synapses) to convey impulses by which one nerve activates another nerve, for sense organs to produce impulses that translate light and sound and touch into electricity, and for nerves and our brains to function. That is, the operation of animal electricity unraveled by Hodgkin and Huxley is what makes it possible for us to read this page, to think about this page, to pick up this page, to call out in surprise, to reflect about *Edge* Questions, and to do everything else that involves motion and sensation and thought. The underlying principle—movement of positively charged particles—was simple, but God resided in the complex details and the elegant reconstruction.

WHY THE GREEKS PAINTED RED PEOPLE ON BLACK POTS

TIMOTHY TAYLOR
Archaeologist, University of Bradford, UK; author, The Artificial Ape

An explanation of something that seems not to need explaining is good. If it leads to further explanations of things that didn't seem to need explaining, that's better. If it makes a massive stink, as academic vested interests attempt to preserve the status quo in the face of far-reaching implications, it is one of the best. I have chosen Michael Vickers's simple and immensely influential explanation of why the ancient Greeks painted little red figures on their pots.

The "red-figure vase" is an icon of antiquity. The phrase is frequently seen on museum labels, and the question of why the figures were not white, yellow, purple, or black—other colors the Greeks could and did produce in pottery slips and glazes—does not seem important. Practically speaking, Greek pottery buyers could mix and match without fear of clashing styles, and the basic scheme allowed the potters to focus on their real passion: narrative storytelling. The black background and red silhouettes make complex scenes—mythological, martial, industrial, domestic, sporting, and ambitiously sexual—graphically crisp. Anyone can understand what is going on (for which reason museums often keep their straight, gay, lesbian, group, bestial, and *olisbos* [dildo-themed] stuff out of public view, in study collections).

Vickers's brilliance was to take an idea well known to the scholar Vitruvius in the first century B.C. and apply it in a fresh context. Vitruvius noted that many features of Greek temples that seemed

merely decorative were a hangover from earlier practical considerations: Little rows of carefully masoned cubes and gaps just under
the roof line were in fact a *skeuomorph*, or formal echo, of the beam
ends and rafters that had projected at that point when the structures were made of wood. Michael argued that Greek pottery
was skeuomorphic too, being the cheap substitute for aristocratic
precious metal. He argued that the red figures on black imitated
gilded figures on silver, while the shapes of the pots, with their
sharp carinations and thin, straplike handles, so easily broken in
clay, were direct translations of the silversmith's craft.

This still seems implausible to many. But to those of us, like
myself, working in the wilds of Eastern European Iron Age
archaeology, with its ostentatious barbarian grave mounds packed
with precious-metal luxuries, it makes perfect sense. Ancient silver
appears black on discovery, and the golden figuration is a strongly
contrasting reddish-gold. Museums typically used to "conserve"
such vessels, not realizing that (as we now know) the sulfidized
burnish to the gold was deliberate and that no Greek would have
been seen dead with shiny silver (a style choice of the hated Per-
sians, who flaunted their access to the exotic lemons with which
they cleaned it).

For me, an enthusiast from the start, the killer moment was
when Vickers photographed a set of *lekythoi*, elegant little cylindri-
cal oil or perfume jars, laid down end to end in decreasing order
in an elegant curve. He demonstrated thereby that no *lekythos* (the
only type of major pottery with a white background, and black
only for base and lid) had a diameter larger than the largest lamel-
lar cylinder that could be obtained from an elephant tusk. These
vessels, he explained, were skeuomorphs of silver-mounted ivory
originals.

The implications are not yet settled, but the reputation of
ancient Greece as a philosophically oriented, art-for-art's sake

culture can now be contrasted with an image of a world where everyone wanted desperately to emulate the wealthy owners of slave-powered silver mines, with their fleets of trade galleys. In my view, the scale of the ancient economy in every dimension—slavery, trade, population levels, social stratification—has been systematically underestimated, and with it the impact of colonialism and emergent social complexity in prehistoric Eurasia.

The irony for the modern art world is that the red-figure vases that change hands for vast sums today are not what the Greeks themselves actually valued. Indeed, it is now clear that the illusion that these intrinsically cheap antiquities were the real McCoy was deliberately fostered, through highly selective use of Greek texts by 19th-century auction houses intent on creating a market.

LANGUAGE AS AN ADAPTIVE SYSTEM

ANDY CLARK

Philosopher, professor of logic and metaphysics, University of Edinburgh; author: Supersizing the Mind: Embodiment, Action, and Cognitive Extension

The iterated-learning explanation of structured language is one of those beautiful explanations that turns things on their head, exposing their workings and origins in a brand-new way. It suggests powerful alternatives to views that depict human brains as heavily adapted to the learning of humanlike languages. Instead, it depicts humanlike languages as heavily adapted to the shape of the learning devices housed in human brains.

The core idea is that language is itself a kind of adaptive system that alters its forms and structures so as to become increasingly easily learnable by the host agents (us). This general idea first appears (as far as I am aware) in Terry Deacon's 1997 book, *The Symbolic Species*. It has been pursued in depth by computationally minded linguists such as Simon Kirby, Morten Christiansen, and others. Much of that work involved computer simulations, but in 2008 Kirby et al. published a paper in *Proceedings of the National Academy of Sciences* augmenting those proofs in principle with a laboratory demonstration using human subjects.

In these experiments, subjects were taught a simple artificial language made up of string/meaning pairs, and then tested on that language. Some of the test items were meanings that had featured in training, while others were new meanings. Then comes the trick. A "new generation" of subjects is then trained, using not the original items but the data from the previous generation. The language is thus forced through a kind of generational bottleneck, such that one generation's choices (including errors and alterations) provide the next generation's data. What the experimenters robustly found (echoing the earlier simulation results) was that languages subject to this kind of cumulative cultural evolution became increasingly easy to learn, exhibiting growing regularities of construction and inflection. This is because the languages alter and morph in ways that are a better and better fit with the basic biases of the subjects (the hosts). In other words, the languages adapt to become easier to learn by the kinds of agent that are there to learn them. They do this because learners' expectations and biases affect both how well they recall actual training items and how they behave when presented with novel ones.

Language thus behaves a bit like an organism adapting to an environmental niche.

We are that niche.

THE MECHANISM OF MEDIOCRITY

NICHOLAS G. CARR

Journalist; author, The Shallows: What the
Internet Is Doing to Our Brains

In 1969, a Canadian-born educator named Laurence J. Peter pricked the maidenhead of American capitalism. "In a hierarchy," he stated, "every employee tends to rise to his level of incompetence." He called it the Peter Principle, and it appeared in a book of the same name. The little volume, not even 180 pages long, went on to become the year's top seller, with some 200,000 copies going out bookstore doors. It's not hard to see why. Not only did the Peter Principle confirm what everyone suspected—bosses are dolts—but it explained why this had to be so. When a person excels at a job, he gets promoted. And he keeps getting promoted until he attains a job that he's not very good at. Then the promotions stop. He has found his level of incompetence. And there he stays, interminably.

The Peter Principle was a hook with many barbs. It didn't just expose the dunderhead in the corner office. It took the centerpiece of the American dream—the desire to climb the ladder of success—and revealed it to be a recipe for mass mediocrity. Enterprise was an elaborate ruse, a vector through which the incompetent made their affliction universal. But there was more. The principle had, as a *New York Times* reviewer put it, "cosmic implications." It wasn't long before scientists developed the "Generalized Peter Principle," which went thus: "In evolution, systems tend to develop up to the limit of their adaptive competence." Everything progresses to the point at which it founders. The shape of existence is the shape of failure.

The most memorable explanations strike us as alarmingly obvious. They take commonplace observations—things we've all experienced—and tease the hidden truth out of them. Most of us go through life bumping into trees. It takes a great explainer, like Laurence J. Peter, to tell us we're in a forest.

THE PRINCIPLE OF EMPIRICISM, OR
SEE FOR YOURSELF

MICHAEL SHERMER

Publisher, Skeptic *magazine; monthly columnist,* Scientific
American; *author,* The Believing Brain

Empiricism is the deepest and broadest principle for explaining the
most phenomena in both the natural and social worlds. Empiricism
is the principle that says we should see for ourselves instead of trusting the authority of others. Empiricism is the foundation of science,
as the motto of the Royal Society of London—the first scientific
institution—notes. *Nullius in Verba—Take nobody's word for it.*

Galileo took nobody's word for it. According to Aristotelian cosmology, the Catholic Church's final and indisputable authority of
Truth on matters heavenly, all objects in space must be perfectly
round and perfectly smooth, and revolve around Earth in perfectly
circular orbits. Yet when Galileo looked for himself through his tiny
tube with a refracting lens on one end and an enlarging eyepiece on
the other, he saw mountains on the moon, spots on the sun, phases
of Venus, moons orbiting Jupiter, and a strange object around Saturn.
Galileo's eminent astronomer colleague at the University of Padua,
Cesare Cremonini, was so committed to Aristotelian cosmology that
he refused even to look through the tube, proclaiming: "I don't believe
that anyone but he saw them, and besides, looking through glasses
would make me dizzy." Those who did look through Galileo's tube
could not believe their eyes—literally. One of Galileo's colleagues
reported that the instrument worked for terrestrial viewing but not
celestial, because "I tested this instrument of Galileo's in a thousand
ways, both on things here below and on those above. Below, it works

wonderfully; in the sky it deceives one."* A professor of mathematics at the Collegio Romano was convinced that Galileo had put the four moons of Jupiter inside the tube. Galileo was apoplectic: "As I wished to show the satellites of Jupiter to the professors in Florence, they would neither see them nor the telescope. These people believe there is no truth to seek in nature, but only in the comparison of texts."†

By looking for themselves, Galileo, Kepler, Newton, and others launched the Scientific Revolution, which in the Enlightenment led scholars to apply the principle of empiricism to the social as well as the natural world. The great political philosopher Thomas Hobbes, for example, fancied himself as the Galileo and William Harvey of society: "Galileus . . . was the first that opened to us the gate of natural philosophy universal, which is the knowledge of the nature of motion. . . . The science of man's body, the most profitable part of natural science, was first discovered with admirable sagacity by our countryman, Doctor Harvey. . . . Natural philosophy is therefore but young; but civil philosophy is yet much younger, as being no older . . . than my own *de Cive*."‡

From the Scientific Revolution through the Enlightenment, the principle of empiricism slowly but ineluctably replaced superstition, dogmatism, and religious authority. Instead of divining truth through the authority of an ancient holy book or philosophical treatise, people began to explore the book of nature for themselves.

Instead of looking at illustrations in illuminated botanical books, scholars went out into nature to see what was actually growing out of the ground.

---

* Daniel J. Boorstin, *The Discoverers* (New York: Random House, 1983), 315-16.

† W. T. Sedgwick & H. W. Tyler, *A Short History of Science* (New York: Macmillan, 1921), 222n.

‡ Hobbes, *De Corpore*, preface to Vol. 1 (1655).

Instead of relying on the woodcuts of dissected bodies in old medical texts, physicians opened bodies themselves to see with their own eyes what was there.

Instead of burning witches after considering the spectral evidence as outlined in the *Malleus Maleficarum*—the authoritative book of witch-hunting—jurists began to consider other forms of more reliable evidence before convicting someone of a crime.

Instead of a tiny handful of elites holding most of the political power by keeping their citizens illiterate, uneducated, and unenlightened, people could see for themselves, through science, literacy, and education, the power and corruption that held them down, and they began to throw off their chains of bondage and demand rights.

Instead of the divine right of kings, people demanded the natural right of democracy. Democratic elections, in this sense, are scientific experiments: Every couple of years, you carefully alter the variables with an election and observe the results. Many of our founding fathers were scientists who deliberately adapted the method of data-gathering, hypothesis-testing, and theory formation to their nation-building. Their understanding of the provisional nature of findings led them to form a social system wherein empiricism was the centerpiece of a functional polity. The new government was like a scientific laboratory, conducting a series of experiments year by year, state by state. The point was not to promote this or that political system but to set up a system whereby people could experiment to see what works. That is the principle of empiricism applied to the social world.

As Thomas Jefferson wrote to John Tyler in 1804: "No experiment can be more interesting than that we are now trying, and which we trust will end in establishing the fact, that man may be governed by reason and truth."

WE ARE STARDUST

KEVIN KELLY

Editor-at-large, Wired; *author,* What Technology Wants

Where did we come from? I find the explanation that we were made in stars to be deep, elegant, and beautiful. This explanation says that most atoms in each of our bodies were built up out of smaller particles produced in the furnaces of long-gone stars. Only our primordial hydrogen bits were born before stars. In a cosmic accounting, we are 90 percent star remnants. At our core, humans are essentially the by-products of nuclear fusion. The intense pressures and temperatures of these giant stoves thickened collapsing clouds of tiny elemental bits into heavier bits, which, once fused, were blown out into space as the furnace died. The heaviest atoms in our bones may have required more than one cycle in the star furnaces to fatten up. Uncountable numbers of built-up atoms congealed into a planet, and a strange disequilibrium called life swept up a subset of those atoms into our mortal selves. We are all collected stardust. And by a most elegant and remarkable transformation, our starstuff is capable of looking into the night sky to perceive other stars shining. They seem remote and distant, but we are really very close to them, no matter how many light-years away they are. All we see of one another was born in stars. How beautiful is that?

INDEX

BOOKS BY JOHN BROCKMAN

ISBN 978-0-06-6223017-1
(paperback)

ISBN 978-0-06-202313-1
(paperback)

ISBN 978-0-06-202584-5
(paperback)

ISBN 978-0-06-210939-2
(paperback)

ISBN 978-0-06-202044-4
(paperback)

ISBN 978-0-06-189967-6
(paperback)

ISBN 978-0-06-143693-2
(paperback)

ISBN 978-0-06-121495-0
(paperback)

ISBN 978-0-06-084181-2
(paperback)

Available wherever books are sold, or call 1-800-331-3761 to order.